The
Clonal Basis of
Development

THE CLONAL BASIS OF DEVELOPMENT

The Thirty-Sixth Symposium of The Society for Developmental Biology

Raleigh, North Carolina, June 13-15, 1977

EXECUTIVE COMMITTEE:

1976-1977

IAN M. SUSSEX, Yale University, *President*
WILLIAM J. RUTTER, University of California, *Past-President*
IRWIN R. KONIGSBERG, University of Virginia, *President-Designate*
WINIFRED W. DOANE, Yale University, *Secretary*
MARIE DI BERARDINO, Medical College of Pennsylvania, *Treasurer*
VIRGINIA WALBOT, Washington University, *Member-at-Large*

1977-1978

IRWIN R. KONIGSBERG, University of Virginia, *President*
IAN M. SUSSEX, Yale University, *Past-President*
NORMAN WESSELLS, Stanford University, *President-Designate*
WINIFRED W. DOANE, Yale University, *Secretary*
MARIE DI BERARDINO, Medical College of Pennsylvania, *Treasurer*
GERALD M. KIDDER, University of Western Ontario, *Member-at-Large*

Business Manager
CLAUDIA FORET
P.O. BOX 43
Eliot, Maine 03903

The
Clonal Basis of
Development

Stephen Subtelny, Editor

Department of Biology
Rice University
Houston, Texas

Ian M. Sussex, Co-Editor

Biology Department
Yale University
New Haven, Connecticut

ACADEMIC PRESS   New York   San Francisco   London   1978
A Subsidiary of Harcourt Brace Jovanovich, Publishers

ACADEMIC PRESS, INC.
111 Fifth Avenue, New York, New York 10003

United Kingdom Edition published by
ACADEMIC PRESS, INC. (LONDON) LTD.
24/28 Oval Road, London NW1 7DX

Library of Congress Cataloging in Publication Data

Society for Developmental Biology.
  The clonal basis of development.

  (Symposium of the Society for Developmental Biology;
36th)
  Proceedings of the symposium held in Raleigh, N.C.,
June 13–15, 1977.
  Includes index.
  1. Developmental biology—Congresses.   2. Clone
cells—Congresses.   I. Subtelny, Stephen Stanley,
Date   II. Sussex, Ian M., Date   III. Title.
IV.   Series: Society for Developmental Biology.
Symposium; 36th. [DNLM: 1.   Clone cells—Congresses.
QH585 S678c]
QH511.S6   no. 36   574.3'08s [574.3]   78-23508
ISBN 0-12-612982-7

Contents

IV. Nuclear and Genetic Events
in Clone Initiation

Preface

The idea that embryos consist of a complex array of morphogenetic fields formed the conceptual basis of developmental biology for many years. Increasingly, however, field theories of development are under attack. They have not provided satisfactory answers to questions of why the field boundaries are so sharp, or why there is no gradient of cell type within the field. And attempts to identify morphogenetic substances that might be the chemical basis of the field have not met with general success.

An alternative view that has emerged is that the early embryo is a cell clone within which subclones differentiate to give rise to specific structures or parts of structures. This view of the developing organism has progressed farthest in *Drosophila* where the ability to mark cell clones by mutations has provided a powerful stimulus to progress, and has extended to mammalian embryological studies with the introduction of chimeric technology. Quite independently botanists had been making chimeric plants since the pioneering work of Winkler in the early 1900s, but because they had been articulating their results in a different terminology the similarities between the plant and animal results had until recently been overlooked. The 36th Symposium of the Society focused on clonal aspects of development, to see where we have come from, where we are now, and where we are going with this approach to developmental analysis. The five sessions examined clonal analysis in *Drosophila,* in mammals, in plants, and in lateral organs of plants and animals, and concluded with an examination of genetic mechanisms of clone initiation.

The Symposium was held on the campus of North Carolina State University, and its success was due in large part to the excellent work of the local committee headed by John Scandalios, and to the efforts of Claudia Foret who was our liaison with the local committee. Financial support provided by the Developmental Program of National Science Foundation made it possible to bring an outstanding group of scientists to speak at the symposium. The Society deeply appreciates the continuing financial support provided to us by the National Science

Foundation, and recognizes the importance of this to progress in the whole field of developmental biology.

With this volume we change editors for the symposium series. John Papaconstantinou, edited volumes 33-35. The Society expresses its thanks to him for his excellent services during this time.

Ian M. Sussex

I. Invertebrates

The Initiation and Maintenance of Gene Activity in a Developmental Pathway of *Drosophila*

A. García-Bellido and M. Paz Capdevila

Centro de Biología Molecular, C.S.I.C.
Universidad Autonóma de Madrid
Madrid, Spain

I. INTRODUCTION

Determination during embryonic development can be considered as a discrete event preceding and leading to terminal differentiation. Whereas differentiation could be molecularly defined by a spectrum of specific cell products, determination is an operational concept with unknown genetic or molecular bases. However, the existence of tissue-specific hormone receptors or of tissue cell lines stable upon transplantation and culture suggests that the final inventory of gene products was somehow already defined at the genetic level. Thus, the problem of cell differentiation is implicit in the problem of determination. This in turn could depend on the mechanism of activating certain genes or groups of genes and maintaining them in an active state in subsequent cell generations.

It was possible to elucidate the mechanism involved in gene activation in microorganisms because the elements involved could be manipulated. The minimum number of elements is two: the specific gene to be activated (structural gene) and an extrinsic factor (inductor)

3

specific for the activation. Since this specificity is probably based on molecular recognition, inductor molecules would presumably first interact with a gene product in order to affect the DNA. Therefore, in the simplest scheme of regulation two genes are involved: a regulator and a structural gene. Very little is known of the mechanisms of gene activation during embryonic development for several reasons: the lack of well defined genetic variants, the lack of knowledge about the specific inductor molecules and the difficulties involved in the manipulation of both in cells.

We will discuss here some data which suggest that regulatory mechanisms function during the determination of the metathoracic developmental pathway in *Drosophila*. This study is based, on the one hand, on the possibility of manipulating different genetic variants that affect this pathway. Since such genetic variants can be defined at the cellular level, their cellular phenotypes are a reliable indication of gene activity. On the other hand, it is based on the possibility of experimentally varying the local distribution of extrinsic factors apparently functioning during the initiation of the pathway.

II. THE GENETIC ELEMENTS

A. *Genetic Interactions*

The development of a normal metathoracic segment requires the function of the wild type alleles of the bithorax gene complex (see Lindsley and Grell, 1968 for details about the genetic variants mentioned). Genetic analysis has shown that this gene complex codes for different complementing functions (Lewis, 1964, 1967), (Fig. 1). It is located in the bands 89E1-4 in salivary chromosomes. Certain mutant alleles in the bithorax system cause transformations between mesotho-

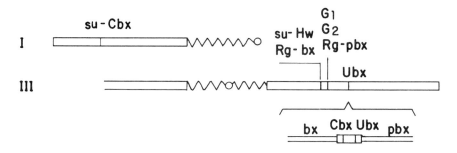

Fig. 1. Distribution in the *Drosophila* genome of the loci involved in the metathoracic pathway. Description of genotypes and mutant interactions in text.

racic and metathoracic segments (Table I). In the adult fly, such transformations are visible in all cuticular derivatives of the meso-thorax and metathorax. Mutant alleles fall into two main groups:

TABLE I

Genetic, Phenotypic, Clonal, and Phenocopy Properties of Different Alleles and Loci Related to the Metathoracic Pathway

| Locus | Allele | HZ | HM | Phenotype in Flies | | | Phenotype in Clones | | Phenocopy | |
				Pene-trance	Expres-sivity	Speci-ficity	Clonality	Fidelity	Via ♀	Via ♂
bithorax	bx³⁴ᵉ	+	bx	T	P	C	P	short	1.0	
	bx³	+	bx	T	T-P	C	T-P	short	0.9	1.0
	pbx	+	pbx	T	T-P	C	T-P	short	1.1	
	Ubx	Ubx	L(bx,pbx)	T	T	C		short	0.9	
	Ubx¹³⁰	Ubx	L(bx,pbx)	T	T	C		short	1.7	
	Cbx	Cbx	Cbx	T	P	C	P		0.9	
	Hm	Cbx	L	T	P	C	T-P		0.6 (xx)	
Rg-pbx	Rg-pbx	pbxᵛ	L(x)	P	P	V	T	long	2.4	2.9
	Rev(G₁)	+	L(x)	+	+	−		long	0.9	
	Rev(G₂)	+	L(x)	+	+	−		long	0.8	
Rg-bx	Rg-bx	bxᵛ	L(x)	P	P	V	T(?)	long	3.0	
	Df(3)red	bxᵛ & pbx	L(x)	P	P	V		C.L	3.2	1.8
su-Cbx	su-Cbx	+	su-Cbx	T	T	C	T	short	1.0	1.3
	Df(1)KA14	+	L	T	T	C		C.L	0.8	
su-Hw	su²-Hw	+	su-bx³	T	P	C			1.0	

HZ: heterozygous; HM: homozygous. Phenotypes; +: wildtype; L: lethal; in parentheses the phenotype of the lethal embryo; (x): apparently normal segmentation. Penetrance and expressivity (T: complete, P: partial) specificity (C: constant; V: variegated or variable). Clonality: penetrance in cells of the same clone. Fidelity: perdurance of the maternal phenotype in mitotic recombination clones. C.L: cell lethal. Phenocopy: data presented as the ratio of phenocopy frequencies of mutant zygotes to wildtype sib flies. Via ♀ or via ♂ : maternal or paternal origin of the mutant in the zygote. (x x) phenocopies only in the notum; the wing to capitellum transformation is not changed by the ether treatment.

recessive and dominant. The recessive alleles show, in homozygous condition, transformation either of the anterior *(bithorax (bx)* alleles) or of the posterior *(postbithorax, (pbx)* alleles) developmental compartments of the metathoracic segment into the corresponding ones of the mesothoracic segment. All the *bx* and *pbx* mutants which have been studied are recessive over a single dose of their wildtype alleles. Mutant alleles in homozygous flies show total penetrance but variable expressivity. Their maximal expressivity is restricted to either the anterior or posterior compartment. In weak, "leaky", mutant alleles the partial expressivity has, however, a constant specificity: they transform specific regions within the compartment affected. It is characteristic of these leaky mutants to vary their expressivity under different conditions of temperature and genetic background (Villee, 1945; Kaufman *et al.*, 1973).

Two types of dominant mutants are known in this system: The first group includes the *Ultrabithorax (Ubx)* alleles. All of these correspond to the lack of function of both bithorax and postbithorax wildtype alleles located in cis-configuration. The same phenotype can be caused by point mutations, mapping between *bx* and *pbx* loci, and also by chromosome rearrangements with breakpoints in 89E1-2 and by chromosome deficiencies for those bands. The phenotype of *Ubx*, or of *Df*(89E1-2) in heterozygous flies with one wildtype dose is a slightly swollen haltere. This phenotype disappears in heterozygous flies with the wildtype system duplicated. Thus, the *Ubx* dominant phenotype corresponds to haplo-insufficiency of the system. *Ubx* is lethal in homozygous flies. This phenotype can be studied in embryos or in mitotic recombination clones (see below) in the adult cuticle. Under these conditions they show the double syndrome of extreme *bx* and *pbx* mutants (Lewis, 1964; Morata and García-Bellido, 1976). Embryos homozygous for DF(89E1-4) show segmental transformations in the epidermis and the nervous system (Lewis, pers. comm.). It is possible that the genetic information of the bithorax system is required for the proper segmental development of all the germ layer derivatives.

The second group of dominant mutants includes the *Contrabithorax (Cbx)* alleles. These are dominant over one or several doses of wildtype bithorax systems. They include point mutations *(Cbx)*, mapping close to *Ubx* and between the loci of *bx* and *pbx*, and a rearrangement *(Haltere mimic, Hm)* with a breakpoint in 89E1-2 (Lewis 1964, and pers. comm). The phenotype of these mutants show the transformation of meso-thoracic structures into metathoracic ones, which is opposite to that caused by *Ubx*. The penetrance of these mutations is complete, but the expressivity is partial, increasing with the number of doses of mutant alleles. The specificity varies between different alleles, but is constant for a given allele.

The effects of the different mutant bithorax alleles suggests that the wildtype function of the bithorax system is to create a metathoracic pathway as an alternative to the mesothoracic one. Genetic analysis of the system suggests that it contains structural loci (bx and pbx) and cis-regulatory loci (Ubx and Cbx). The genetic behaviour of the Ubx alleles indicates that they correspond to operator deficient mutations (0°) whereas Cbx alleles can be interpreted as operator constitutive (0^{c}) mutations (Lews, 1964, 1967). The phenotypic transformations caused by Cbx alleles suggests that bithorax system wildtype products are released in the mesothorax. This leads to the suppression of the mesothoracic pathway and the appearance of its metathoracic alternative. That this is due to the derepression of the bithorax system is confirmed by the cis-suppression effect of bx^3 on Cbx in bx^3 Cbx/+ + flies (Lewis, 1964). That Cbx is also derepressed in other segments, is strongly suggested by the effect of Cbx in double mutant combinations with other homeotic mutations which transform other segments to mesothorax. For example, head structures transformed into dorsal mesothoracic structures in Opthalmoptera (Opt) flies (Goldschmidt and Lederman-Klein, 1958) is further converted into metathoracic structures in Opt; Hm flies (Capdevila, unpublished).

The existence of mutations, mapping outside the bithorax system, which show a bithorax phenotype on their own or that interact with the expression of the bithorax mutant alleles, suggests that the metathoracic developmental pathway requires the normal function of other genes besides those of the bithorax system. We will summarize some genetic data relating to these mutants. (Table I, Fig. 1).

The mutant called Regulator of postbithorax (Rg-pbx, Lewis, 1968) is a recessive lethal which shows a dominant visible phenotype over its wildtype allele. The penetrance in heterozygotes is not complete but high (ca. 90%) as is its expressivity. In addition its specificity is variable; heterozygous flies show, in an asymmetric and erratic fashion, mesothoracic transformation patches. These appear anywhere within the posterior compartment of the metathoracic segment. The mutant condition is associated with one of the breakpoints of the In(3R) 85B; 88B. A duplication carrying the left hand region of this inversion on the Y chromosome shows the same dominant phenotype as the original inversion. Penetrance and expressivity are similar in zygotes receiving the mutation from either parental gamete. Penetrance and expressivity vary with the number of wildtype doses of the bithorax system present in the genome; both decrease with increasing number of these wildtype genes (Lewis, 1967). This behavior led E. B. Lewis to suggest that the mutation may lead to a superrepressor condition in a regulator locus with trans effects on the *pbx* gene. It is therefore interesting to note that the mutant phenotype is suppressed in *Rg-bx* +/+ *Cbx* flies (Capdevila, unpublished). Two independent revertants (G_1 and G_2) of the *Rg-pbx* original inversion have been isolated by E. B. Lewis. Both are homozygous lethal and lethal in heterozygous combination with *Rg-pbx*. It is reasonable to assume that the reversion of the phenotype is associated with the lack of function of the *Rg-pbx* gene. Thus, if the *Rg-pbx* mutation corresponds to a mutation of the trans-regulator gene, its amorphic condition would lead to an extreme phenotype in homozygous flies. However, both the *Rg-pbx* and the two revertant mutations in homozygous mutant embryos do not show segmental transformations. Although if only specific for the *pbx+* function, they might not be detectable in cuticular structures.

The point mutant named *Regulator* of *bithorax* (*Rg-bx*, Lewis, pers. comm.) is included in a deficiency, *Df(3)red* (88B), which lacks two or three bands including the loci of *red* and *su-Hw* (see below). Both *Df(3)red* and *Rg-bx* are recessive lethals. Both show, in heterozygous condition over their wildtype homologs, a low penetrance and expressivity. The phenotype is expressed in patches of mesothoracic structures, located in variegated or random fashion (variable specificity), in both anterior and posterior compartments of the metathorax. Thus, the phenotype of *Rg-bx* seems to correspond to the haplo-insufficiency of this locus. Although *Df(3)red* and *In(3)Rg-pbx* have breakpoints in the same chromosome region, they are not allelic. Both *Df(3)red* and *Rg-bx* mutants are viable in flies doubly heterozygous for either *Rg-pbx* or the G_1 or G_2 revertants.

The penetrance of the transformation varies depending on whether the mutation was carried to the zygote by the male or by the female gamete. The penetrance also varies depending on the genetic constitution of the zygote with respect to the *bithorax* system and the *Rg-pbx* locus (Capdevila, 1977). This penetrance is 4% in *Df(3)red*/+ flies of mutant mothers but zero in mutant zygotes of mutant fathers and in non-mutant zygotes of mutant mothers. When the zygotes are also double heterozygotes for *In(3)Ubx¹³⁰* (or *Df(3)89E1-4*) the penetrance increases to about 30% if *Df(3)red* was carried by the female gamete and to about 4% if carried by the sperm. The doubly heterozygous *Df(3)red* +/+ *Rg-pbx* zygotes of *Df(3)red*/+ mothers show the typical high penetrance of posterior transformation caused by *Rg-pbx* but the penetrance of anterior transformation is increased to 30%. These results suggest that the haplo-insufficiency of $Df(3)red$ is expressed in the oocyte. They also indicate that the genes $Rg\text{-}bx$, $Rg\text{-}pbx$ and bithorax have synergistic effects. The lethal homozygous $Df(3)red$ and $Rg\text{-}bx$ embryos do not show noticeable segmental transformations.

Among the various mutants that modify the penetrance and expressivity of mutations in the bithorax system, only those with specific interactions will be discussed. The mutant $su^2\text{-}Hw$ (included in the $Df(3)red$) when in homozygous condition suppresses the pheno-type of certain alleles of the bithorax system: bx^3, bx^{34e}, but neither bx, Cbx, pbx nor Ubx (Lewis, 1967). This allele specific but not locus specific effect suggests that $su^2\text{-}Hw$ may correspond to a translational suppressor (Lewis, 1967). The mutant $su^2\text{-}Hw$ complements for viability and suppression with the $Rg\text{-}bx$ point mutation, implying they are not allelic.

Another mutant, $su\text{-}Cbx$, mapping in the first chromosome, shows specific suppression of Cbx genotypes: it suppresses Cbx but not Hm. This mutant is included in the $Df(1)KA14$ (7F1-2; 8C6; Lefevre, pers. comm.). The suppression of the Cbx phenotype is more effective in $su\text{-}Cbx/Df(1)KA14;Cbx/+$ flies than in $su\text{-}Cbx/su\text{-}Cbx; Cbx/+$ flies. This genetic behaviour suggests that $su\text{-}Cbx$ is a hypomorphic mutation. In these flies the phenotype corresponds to the reversal of the Cbx transformation in the mesothorax. However, $su\text{-}Cbx; Cbx/Ubx$ flies show, besides this reversal, bx and pbx transformations in the meso-thorax. In addition, $su\text{-}Cbx/Df(1)KA14;Cbx/Ubx$ flies show a total mesothoracic transformation of the metathorax; surprisingly $su\text{-}Cbx/Df(1)KA14;Ubx/+$ flies are normal indicating that $su\text{-}Cbx$ (Capdevila, 1977) specifically represses the Cbx chromosome. Note in this context that the pbx phenotype which was suppressed in $Rg\text{-}pbx +/+ Cbx$ flies is fully expressed in $su\text{-}Cbx; Rg\text{-}pbx/+ Cbx$ flies. The genetic behaviour discussed above suggests that the wildtype allele of $su\text{-}Cbx$ is somehow involved in the process of activation of the bithorax system (see discussion).

B. *Clonal Analysis*

Clones of homozygous cells in heterozygous individuals can be generated by mitotic recombination induced at different develop-mental stages. The study of these clones allows us to investigate the degree of cell autonomy and the developmental stage of gene expression. Two main approaches have been used in clonal analysis of morpho-genetic mutants. In the first, "cell lineage analysis", the cell proliferation and expressivity of the mutants are studied in clones expressing cell marker mutants in flies which contain the morphogenetic mutant of interest. In the second approach, "morphogenetic mosaicism analysis", the clonal effects of changes in the genetic constitution with respect to the morphogenetic mutant can be studied in clones initiated at different developmental stages (Fig. 2). If the morphogenetic mutant is recessive, the phenotype of homozygous mutant clones can be analyzed. If dominant, the phenotype of the resulting homozygous mutant or homozygous wildtype cells can similarly be studied.

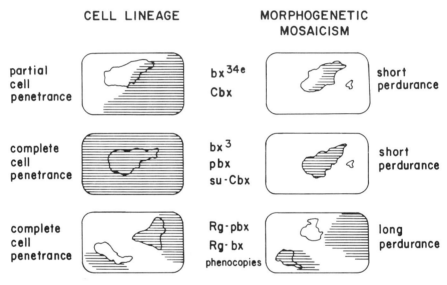

Fig. 2. Clonality of different types of mutations and phenocopies in cell lineage analysis or in morphogenetic mosaics. Areas correspond to segments with complete or incomplete transformations (dashed regions). Outlines of patches correspond to clones. In cell lineage experiments those transformations can be expressed in all the cells or only in some cells within a clone (cell penetrance and clonality of the transformation). In morphogenetic mosaics the allelic substitutions caused by mitotic recombination might lead to a phenotypic transformation in clones (bx^{34e}, bx^3, pbx) or not (Rg-pbx or Rg-bx).

Cell lineage analysis of the bithorax mutants has shown that in mutants like bx^3 or *pbx*, which cause an almost total transformation, clones of marked cells contain only transformed wing cells (Morata and García-Bellido, 1976). In mutants with incomplete expressivity, like bx^{34e}, clones will probably include transformed wing and untransformed haltere cells, indicating that the cell penetrance of the mutation is not clonally determined. Similarly, the transformation caused by *Cbx* in mesothorax is not clonal, suggesting that the variable phenotype results from variable amounts of *bx* and *pbx* wildtype products during successive cell proliferation cycles (Morata, 1975). In morpho-genetic mosaics, the bx^3, *pbx* and *Ubx* homozygous cells show a complete cell autonomy of the transformation, corresponding to the total expressivity of the transformation in homozygous flies (Morata and García-Bellido, 1976). However, the transformation is only expressed in clones larger than 20 cells; smaller clones, resulting from 2-3 cell divisions, remain untransformed, probably due to the perdurance of the bithorax wildtype products (Morata and García-Bellido, 1976). In conclusion, the bithorax wildtype function is required throughout the proliferative divisions of the disc in order to maintain the meta-thoracic pathway.

The variable expressivity of the transformation caused by *Rg-pbx* is not due to variable cell penetrance of the mutant. Cell lineage analysis of *Rg-pbx*/+ halteres has shown that the variegated pheno-type is clonal: clones initiated during the proliferation period are restricted to either transformed (wing) or non-transformed (haltere) territories. The behavior of the *Rg-pbx* mutation in morphogenetic mosaics is also different from that shown by mutants in the bithorax system. In heterozygous *Rg-pbx* halteres, large clones of cells carrying markers that indicate either the removal of the mutant condition or of the wildtype allele, retain their original tissue phenotype: that is, they remain either mesothoracic or metathoracic (Fig. 2) (Capdevila, 1977). Thus, the *Rg-pbx* effect has a long perdurance, possibly from the initiation of the haltere development. Interestingly, the phenotype of G_1 or G_2 homozygous cells in mitotic recombination clones is also that typical of the segment in which they appear: they are metathoracic in the metathorax and mesothoracic in the mesothorax (Capdevila, 1977). It is not yet known what the phenotype of these clones is if they are initiated before the blastoderm stage. The patchy and variable specificity of the transformation indicates that the effect of the *Rg-pbx* mutation occurs at random in the cell population of the meta-thoracic anlage.

The low penetrance of the transformation caused by *Df(3)red* or *Rg-bx* in heterozygous flies does not allow cell lineage analysis. The patchy and random location of the transformation suggests, however, a clonal origin. *Df(3)red* is cell lethal in homozygous mitotic recombina-tion clones, but *Rg-bx* is not. Clones of cells of the latter genotype do not show transformation in either the meta or mesothorax suggesting that, as with *Rg-pbx*, the wildtype allele has a long perdurance.

The lack of clonal expressivity of *Cbx* is compatible with the absence of clonal effects in the suppression of *Cbx* in *su-Cbx; Cbx* flies

(Capdevila, 1977). However, clones of *su-Cbx* cells in *Cbx* flies show cell autonomous suppression. This indicates that the *su-Cbx* function is required throughout development. Clonal analysis has, so far, not been carried out in su^2-*Hw* flies.

III. THE EPIGENETIC DETERMINANTS

External agents applied during early stages of development may affect the development of mesothoracic or metathoracic pathways. Thus, the exposure of wildtype embryos to temperature shock (Maas, 1948) or to ether vapour (Gloor, 1947) produces phenocopies of the bithorax mutants. Most of the phenocopies correspond to the transformation caused by *bithorax* (ca. 90%) but *postbithorax, Contrabithorax* and abdominal transformations also occur in lower frequencies (Capdevila and García-Bellido, 1974, 1977). The effective period for inducing these phenocopies is during the first 4 hours of development. For those caused by ether the maximal response is at about 2-1/2 hours, corresponding to the blastoderm stage (Capdevila and García-Bellido, 1974). When treated at this stage 30% of the adults show phenocopies. In these phenocopies, the transformation can affect the entire anterior metathoracic compartment, as represented in the adult cuticle. In most cases, however, the transformation is patchy and random in position in the metathoracic cuticular derivatives. The frequency of double-sided phenocopies is much higher than expected if they resulted from independent events in the left and right imaginal disc anlagen (Gloor, 1947; Capdevila and García-Bellido, 1974). This high left-right correlation remains when only one side of the embryo is exposed to ether vapour. Ligation experiments in preblastoderm embryos of the dipteran *Smithia* (Herth and Sander, 1973) and *Drosophila* (Schubiger, 1976) have shown that the segmental organization is not laid down in the unfertilized egg, but results from regulative mechanisms in the preblastoderm embryo. The possibility of obtaining phenocopies by treatment of preblastoderm embryos and the high left-right correlation suggest that the ether may interfere with the mechanism of coordinated segmental organization.

Several arguments indicate that the effect of ether is registered at the cellular level in the blastoderm and maintained clonally in successive cell generations. That phenocopy patches are of clonal origin is suggested by their shape and sharp borders, their random position, and because they respect compartment boundaries. All these characteristics are similar to those of mitotic recombination clones. The clonal origin of phenocopied patches is demonstrated in cell lineage analyses. Clones initiated after 48 hours of development are restricted to either the transformed or non-transformed territories running for hundreds of cells along the boundary between meso and meta-thoracic tissue (Fig. 2) (Capdevila and García-Bellido, 1974). The lasting effect of the ether treatment upon the progeny of the blastoderm cells is incompatible with the idea that phenocopies result from depletion of wildtype products of the bithorax system. These are known to be required and produced during successive cell generations. Therefore, the phenocopying event may result from perturbing the mechanism of activation of the metathoracic pathway in individual cells. Such a mechanism requires the existence of a positional signal within the egg and its registering and expression, in genetic terms, in individual cells. If ether treatment perturbs the positional signals, then this can be used as an approach to investigate the mechanism of registering such signals.

We will now discuss the effect of the genetic constitution of the embryonic cells on their sensitivity to ether. This assumes that mutants of genes which are necessary for the registering mechanism may affect the frequency of phenocopies, whereas those of genes necessary for the maintenance of the pathway once initiated will not. Females heterozygous for different mutants were crossed to wildtype standard males and their offspring exposed to ether treatment. In this way heterozygous mutant and wildtype sib embryos were treated under the same genetic and physiological conditions.

The study of embryos heterozygous for the mutants of the struc-tural genes of bithorax (bx, bx³, pbx, bxd) shows that they have the same sensitivity as sib control embryos (Table I) (Capdevila and García-Bellido, 1974, 1977). This finding confirms the inference that the ether treatment does not affect processes that depend on the bithorax products. It also demonstrates that phenocopies are not somatic mutations of these genes. The study of the sensitivity of embryos heterozygous for the dominant alleles of the bithorax system has uncovered several relevant features. The frequency of ether induced phenocopies in Ubx'/+ embryos is similar to controls. Note that in bx³ pbx/Ubx' mutant transformed flies, the ether treatment does not rescue the mutant phenotype in the metathorax, nor causes Cbx phenocopies in the mesothorax. Cbx/+ embryos, on the other hand, show normal frequencies of bithorax phenocopies in the metathorax and (10%) show reversion of the metathoracic transformation in the mesothorax (Capdevila and García-Bellido, 1974, 1977). In conclusion these results indicate that the ether treatment may lead to a repression of the bithorax system in either the metathorax (*bx* and *pbx* phenocopies) or in the mesothorax (*Cbx* reversion) and can also lead to its derepression in the mesothorax (*Cbx* phenocopies). Ether treatment, however, cannot initiate or create a metathoracic pathway in flies mutant for the bithorax structural genes. This implies that the metathoracic pathway has no alternative at the step controlled by the bithorax genes.

The response to ether of two rearrangements with breakpoints in the bithorax system (*Ubx*[130] and *Hm*) is different to that of point mutants of the same loci (*Ubx'* and *Cbx*). The *Hm* phenotype cannot be reverted, nor do wing transformations occur, in the metathorax of such flies. Thus, contrary to the *Cbx* behaviour that of *Hm* is consistent with the interpretation that this mutation is constitutive and non-repressible. Heterozygous *Ubx*[130] embryos, on the other hand, show twice as many phenocopies as controls. Rearrangements with breakpoints near the locus of the bithorax system are known to interact with the degree of expression of the bithorax mutants located in trans configuration, ("transvection" effects, Lewis, 1954). We therefore studied the effect of two rearrangements with breakpoints near bithorax, *In(3LR)TM1* and *T(2;3)apXa*. Both show the same sensitivity to phenocopies as control embryos. Thus the different behaviour of *Ubx*[130], compared to *Ubx'*, could result from a structural alteration in the bithorax region. We would then expect that deficiencies for the bithorax region should have the same effect as the *Ubx*[130] chromosome on the sensitivity to phenocopies conferred to heterozygous embryos. Two small deficiencies lacking bands 8E1-4, ie. the entire bithorax system *(Df(3)P9* and *Df(3)P115)* double the frequency of phenocopies in heterozygous embryos. Conversely, embryos carrying three or four doses of these regions, somehwere in the genome, showed respectively about one-half and one-fourth the sensitivity of controls (Capdevila, 1977; Capdevila and García-Bellido, 1977).

This result is meaningful in several respects: i) it indicates that the clonal effects of the ether do not result from replicative modification of any particular cytoplasmic factor. ii) the element which registers the alteration is in the bithorax system. iii) since increasing doses of the bithorax genes lower the response to ether-induced phenocopies these must result from repression of this system.

IV. THE ACTIVATION MECHANISM

The above results suggest that the bithorax system is the genetic element that registers the alteration caused by ether treatment. They further indicate that the functioning or non-functioning of the bithorax system depends on a negative type of control. The expression of bithorax in the metathorax requires that at least one copy of this gene per cell is not repressed. We postulate that in the untreated embryo the bithorax genes become active in the cells of the metathoracic region because a positional signal in this region prevents the bithorax genes from being repressed (Fig. 3). In the mesothoracic segment such a signal is either absent, or so weak so, that the bithorax genes become repressed. The disruptive effect of ether might consist in shifting the boundary of such a signal, so that some cells in an otherwise meta-thoracic position become depleted of this signal and develop auto-nomously as mesothoracic cells (*bx* or *pbx* phenocopies). Alternatively, cells in an otherwise mesothoracic position may have enough signal to prevent bithorax genes from repression giving rise to the clones of *Cbx* phenocopies (Fig. 3) (Capdevila, 1977).

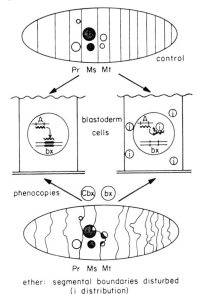

Fig. 3. Hypothetical normal and ether disturbed mechanism of the initiation of mesothoracic and metathoracic pathways at blastoderm. Pr, MS, MT: pro, meso and metathoracic segments; A: repressor coding gene; i: inductor or positional signal; bx: bithorax system. *Cbx* and *bx*: mosaic meso and meta-thoracic anlagen giving rise to the corresponding (circle) phenocopies.

The operational similarity between the positional signal and the inductor or effector molecule in a system of control of genetic expression by repression is obvious. In such a system the repression results from the interaction of 1) an inductor molecule, 2) a cis-regulatory region in the structural gene and 3) the product of a trans-regulatory gene. If a similar system is at work in the control of the initiation of the metathoracic pathway, inductor molecule(s) may be represented by the positional signals and the structural gene by the bithorax system. It is not known whether these positional signals are diffusible molecules. Neither is it known that genes responsible for their syntheses nor for the synthesis of the repressor molecules exist. However, if they are direct or indirect gene products it should be possible to identify such genes by studying the effects of mutants upon the sensitivity to phenocopies.

We have studied the sensitivity to phenocopies of embryos heterozygous for the mutants mapping outside the bithorax system, which either show bithorax phenotypes or modify the phenotype of bithorax mutants. Mutants like su-Cbx, its deficiency and su-Hw do not change the responsiveness to ether (Table I). This result is consistent with previous genetic and developmental data that suggested that their effects were on the products of the bithorax mutants during the proliferation phase of development.

A different situation was found with the mutants Rg-pbx and Rg-bx. Heterozygous Rg-pbx embryos show a higher frequency of phenocopies than their controls (Table I). Moreover, the frequency of phenocopies in zygotes carying the Rg-pbx mutation is similar irrespective of whether the mutation came through the maternal or paternal gamete. This higher ether sensitivity of Rg-pbx embryos is consistent the hypothesis that the mutant condition results in a superrepressor molecule that escapes the inductor and binds to the cis-regulatory region of the bithorax system. Embryos heterozygous for either the G_1 or G_2 mutation show the same sensitivity to phenocopies as sib-controls. This would be expected on the hypothesis that the amount of repressor molecules produced by a single locus is sufficient for repression. Note, however, that the frequency of Cbx phenocopies in these embryos is also normal.

As mentioned above the penetrance of the Df(3)red phenotype is higher in embryos receiving the mutation from the mother as opposed to the father. Events are similar with respect to the frequency of phenocopies. The same was found with the point mutant Rg-bx. The increase in frequency of phenocopies in Df(3)red or Rg-bx embryos is

3 times that of control embryos. The increase in phenocopy frequency, the maternal effects and the fact that the phenotype corresponds to the haplo-insufficiency of the locus are all consistent with the normal function of this gene being involved with either the synthesis or the distribution of inductor molecules.

The preceding genetic, developmental and phenocopy data are consistent with a mechanism of initiation requiring at least three elements: 1) the bithorax system, 2) a mutant *Rg-pbx* whose product shows properties functionally similar to those of a superrepressor molecule, and 3) a positional inductor, the synthesis or distribution of which requires the *Rg-bx* gene. It is the distribution of this inductor within the egg cortex which apparently is altered by ether treatment.

If, as a result of the ether treatment, the concentration of a positional inductor in the presumptive metathoracic segment is depleted, an excess of repressor molecules will be available to bind the bithorax genes. The haplo-insufficiency of *Df(3)red*, or *Rg-bx* in heterozygous embryos has synergistic effects with the ether treatment. This is possibly because both cause depletion of positional inductors in the metathorax. The patchy appearance of the bithorax transformation of *Df(3)red* or *Rg-bx* is consistent with the hypothesis that they are caused by cell autonomous failures in the initiation of the metathoracic pathway.

If the *Rg-bx* gene is not directly involved in the synthesis of the particular positional inductor, mutants at other loci would be expected to show similar genetic behavior and sensitivity to ether phenocopies as that of *Rg-bx* mutants.

The effect of the mutation *Rg-pbx* in the production of bithorax phenocopies can be explained under the assumption that it corresponds to a superrepressor modification of the normal repressor molecule: this would occassionally bind its receptor site on the bithorax gene even in the presence of inductor and more efficiently so in its absence. However, several arguments suggest that the *Rg-pbx* gene does not correspond to the gene coding for the normal repressor: 1) the lack of effect of its revertants (G_1 and G_2) in increasing the frequency of *Cbx* phenocopies, and 2) the lack of morphologically visible homoeotic transformation in the lethal G_1/G_2 embryos. The phenotypic effect of *Rg-pbx* might result from a modified gene product which happens to have an affinity for the repressor binding site of the bithorax system.

Presumably, the repressor binding site extends over a specific site in the bithorax system. The higher frequency of phenocopies in *In(3)Ubx*[130] and *Df(3)P115* heterozygous embryos suggests that they result from the incapacity of a receptor site to bind the repressor. The excess repressor would then be available to bind the receptor binding site in the wildtype homologous chromosome. This interpretation suggests merely that the receptor site is the *Ubx* locus itself. Deficiencies and duplications of parts of the system were studied to investigate their effects on repressor binding. The duplication *(Dp(3;3)bxd[100])* which carries an extra dose of the *Ubx* locus reduces the frequency of phenocopies.

The observation that phenocopies of *postbithorax* are rare, even in cases where the entire anterior compartment shows a *bithorax* phenocopy, suggests that the *pbx* locus may have a different repressor binding site. The effect of *Rg-pbx* mutant which in untreated embryos is exclusively on the posterior compartment also suggests the existence of an independent control for the *pbx* locus. It is not known whether the repressor binding site for activation of the system and that for initiation of transcription are the same or different sites.

V. THE MAINTENANCE MECHANISM

The results of the genetic, clonal and phenocopy analyses support the interpretation that both *Rg-pbx* and *Rg-bx* affect only the initiation step of the metathoracic or mesothoracic pathways. In fact, as seen above, *Rg-pbx* (and possibly *Rg-bx*) phenotypes have a clonal origin. The genetic change with respect to these alleles is not followed by a change in phenotype in mitotic recombination clones initiated during the proliferation phase. However, the removal of the wildtype alleles of the bithorax system in heterozygous mutant metathoracic cells, during the proliferation phase, is followed by reversion to the mesothoracic phenotype. Thus, the bithorax genes are responsible for the maintenance of the metathoracic pathway. Therefore, the initiation mechanism and the maintenance mechanism are different. Once the bithorax system has been derepressed or repressed at initiation, it remains in a transcribable or non-transcribable condition during successive cell generations. In the derepressed state, it is the allelic condition of the bithorax structural alleles which determines the final phenotype.

A stable state can be maintained in several ways, for example, by structural modifications at the chromatin level. If the repressed state replicates autonomously for each homologous chromosome, it would be possible to produce phenocopy spots by mitotic recombination in cells where only one homolog has been repressed following ether treatment. This could result in clones of homozygous repressed chromosomes. In halteres showing phenocopies, clones of a cell marker, located in the same arm as the bithorax system, were never associated with a transformation in the non-phenocopied areas (Capdevila, unpublished). The negative result of this experiment, in itself, is not conclusive. However, this hypothesis is improbable because of the following consideration. We have seen that hetero-zygous embryos for mutants in the structural loci have the same frequency of phenocopies as control embryos. This could not be the case if, in cells with one single chromosome repressed, this repression had occurred in the non-mutant chromosome. In such cases the resulting clone would have shown a mutant phenotype and therefore the frequency of ether treated embryos with transformed patches increased. We therefore conclude that, if even one bithorax gene of the zygote is not repressed at initiation, then this non-repressed copy will "rescue" the repressed ones in the next cell generation.

Nuclear transplantation experiments (see Gurdon, 1974) and the occurrence of transdetermination (Hadorn, 1966) indicate that cell differentiation can be changed under conditions that affect nucleo-cytoplasmic interactions. It is thus possible that the maintenence of a differentiated state occurs as a result of *reiterative* repression or derepression loops in successive cell generations. If this is the case in the maintenance of the metathoracic or mesothoracic pathways, it is possible that the mesothoracic repression is maintained by the continuous supply of the same repressor molecules operating at the initiation step. The same repressor molecules must thus be inactivated in the metathoracic cells. It is improbable that the synthesis of the inductor, active in the blastoderm, is maintained during the proliferation phase of the metathoracic disc. More likely, the repressor molecules are prevented from binding to the bithorax system by some direct or indirect product of the bithorax system itself. It is interesting to recall, in this context, the interaction between the *Cbx* mutant chromosome and the mutant condition in the *su-Cbx* locus: when both are mutant the bithorax system becomes inactive. Possibly, the maintenance of the derepressed condition of the bithorax system may require one product of the bithorax system which will act as anti-repressor, or inactivator, of the repressor gene (Fig. 4).

DEVELOPMENTAL   PATHWAYS

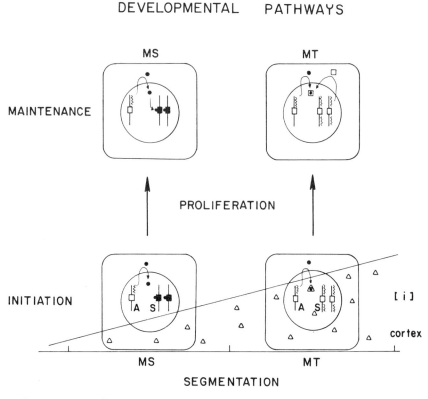

Fig. 4. Genetic mechanisms involved in the initiation step and during maintenance of the meta-thoracic and mesothoracic pathways in proliferating cells. A: regulator gene coding for repressor (•) molecules; S: structural gene, *bithorax* complex, coding for metathoracic characteristics as well as for an antirepressor (□). i: inductor or positional molecules (Δ) which form a concentration gradient in the egg periplasm or cortex. Cis-control region of genes in squares. Arrows: transcription-translation-repression steps.

This model predicts that homozygosis of amorphic alleles of the repressor coding gene, following mitotic recombination during the proliferation of the mesothorax, will show in clones a contrabithorax transformation. It also predicts that it might be possible to experimentally interfere with the interactions between the repressor gene and the bithorax system at any stage during the proliferation phase. The result would be to produce bithorax phenocopies which would also be maintained clonally. This could be the underlying mechanism of the spontaneously occurring trans-determination of metathoracic to mesothoracic pathways (Gehring, *et al.,* 1968).

ACKNOWLEDGMENTS

The authors are thankful to Mrs. P. Garrido and A. C. Andreu for the skilled assistance during the preparation of the unpublished work. Dr. Kankel's critical reading of the MS is most appreciated. This work was supported by grants of the CADC and CAICT.

REFERENCES

Capdevila, M.P. (1977). PhD Thesis, University of Madrid
Capdevila, M.P. and García-Bellido, A. (1974). Nature **250,** 500-502.
Capdevila, M.P. and García-Bellido, A. (1977). (submitted).
Gehring, W., Mindek, G. and Hadorn, E. (1968). J. Embryol. Exp. Morphol. **20,** 307-318.
Gloor, H. (1947). Rev. Suisse. Zool. **54,** 637-712.
Goldschmidt, E. and Lederman-Klein, A. (1958). J. Hered. **49,** 262-266.
Gurdon, J.B. (1974). "The control of gene expression in animal development". Harvard University Press, Cambridge.
Hardorn, E. (1966). Develop. Biol. **13,** 424-509.
Herth, W. and Sander, K. (1973). Wilhelm Roux' Arch. **172,** 1-27.
Kaufman, T.C., Tasaka, S.E. and Suzuki, D.T. (1973). Genetics **75,** 299-321.
Kauffman, S.A. (1973). Science **181,** 310-318.
Lewis, E.B. (1951). Am. Nat. **88,** 225-239.
Lewis, E.B. (1964). In: "Role of chromosomes in development" (M. Locke, ed.), pp. 231-252. Academic Press, New York.
Lewis, E.B. (1967). In: "Heritage from Mendel" (R.A. Brink, ed.) pp. 17-47. University Wisonsin Press, Madison.
Lewis, E.B. (1968). Proc. 12th Int. Congr. Genet. Vol I, 96-97.
Lindsley, D.L. and Grell, E.H. (1968). *Carnegie Inst. Wash. Publ.* No. 627.
Maas, A. (1948). Wilhelm Roux' Arch. **143,** 515-572.
Morata, G. (1975). J. Embryol. Exp. Morphol. **34,** 19-31.
Morata, G. and García-Bellido, A. (1976). Wilhelm Roux' Arch. **179,** 125-143.
Schubiger, G. (1976). Develop. Biol. **50,** 476-488.
Villee, C. (1945). Am. Nat. **79,** 246-258.

The Use of Mosaics to Study Oogenesis
in *Drosophila melanogaster*[*]

Eric Wieschaus

Zoologisches Institut der Universität Zurich
Kunstlergasse 16, 8006 Zurich, Switzerland

I. INTRODUCTION

Drosophila is particularly well suited for a genetic analysis of oogenesis, and a large number of mutations have been isolated which affect this process (Kaplan *et al.*, 1970; Bakken, 1973; Rice, 1973; Gans *et al.*, 1975; Mohler, 1977; reviewed by King and Mohler, 1975). The X-chromosome has been the subject of especially intensive mutagenic screens. Calculations from the Poisson indicate that mutations have been isolated for more than half of the X-linked loci which have their principal effects on oogenesis. All of these mutations depend on the diploid genotype of the mother, and thus transcription occurs during oogenesis

---

[*]Dedicated to Professor Donald Poulson

prior to meiosis. They are called "female sterile mutations" or "maternal effect mutations." Although all involve processes which occur during oogenesis, they can be grouped into three arbitrary classes according to the stage when the defects become noticeable to the experimenter (Gans et al., 1975). Females homozygous for the first class of mutations lay no eggs at all, and histological preparations of ovaries often reveal abnormalities at specific stages of oogenesis. Many mutations are also obtained where eggs are laid but show abnormalities in size or shape, turgor, or the morphology of the protective coverings. Lastly, there are mutations which result in eggs with no visible defects, but which cannot support the normal development of the embryo. The wide variety of defects in all these classes suggests such mutations can be used as probes to define different steps or functions necessary for normal oogenesis. Oogenesis however involves the contributions of several different cell types, and a mere cataloguing of mutant defects provides an extremely inadequate description of the process. In the first half of this report I will describe a number of techniques for making genetic mosaics which can be used to determine the tissue in which a given mutation exerts its primary effect on oogenesis. In the second half, we will use these same mosaic techniques to approach more general questions about the development of the ovary and its function in the adult.

II. THE ROLE OF THE GERMLINE AND SOMATIC TISSUES IN OOGENESIS

In the ovary, we can distinguish two major cell groups which contribute to the developing oocyte. These are the germline derived oocyte nurse cell complex, and the somatically derived follicle cells. Before approaching the question of the role of these cell types during oogenesis, a brief description of the structure of the ovary and the origin of the oocyte is necessary (Fig. 1) (Bucher, 1957; Koch and King, 1966; King, 1970; Mahowald, 1972). The Drosophila ovary is divided into tube-like structures called ovarioles. At the apical end of each ovariole is a germarium which contains the stem populations for both the germline and somatic components of the follicle. After each stem cell division in the germline, one daughter cell divides four more times to produce a cluster of sixteen interconnected cells, one of which will become the oocyte; the others, the nurse cells. This cell cluster is then surrounded by follicle cells whereby it leaves the germarium and enters the main body of the ovariole (the vitellarium). Here it grows and vitellogenesis occurs as it moves down the length of the ovariole. New oocyte cysts are formed

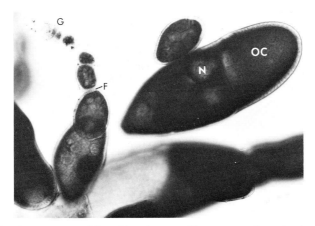

Fig. 1. Developing oocytes of *Drosophila melanogaster*. The oocytes and associated nurse cells are grouped in tube-like structures called ovarioles. The oocyte cysts are formed at the apical end of each ovariole in the germarium (G). In the main body of the ovariole they are arranged in order of advancement with the oldest in the most basal positions. The preparation has been stained for aldehyde oxidase activity (Janning, 1976) and reveals the normal pattern found in wild-type non-mosaic ovaries. The derivatives of the germline, i.e., the oocyte and nurse cells, stain darkly, whereas little or no activity is found in the follicle cells of somatic origin. OC, oocyte; N., nurse cell; F, follicle cell epithelium.

continually at the apical end as mature eggs leave the ovariole basally. Each ovariole contains about 6-7 oocyte cysts, arranged in chronological order with the youngest and smallest cysts in the most apical positions.

Our understanding of the roles of the different cell types in oogenesis is based primarily on morphological observations in *Drosophila* and other insects, coupled with some studies on the incorporation of labelled precursors for DNA, RNA, and protein (King and Burnett, 1959; Zalokar, 1960; Bier, 1963; Mahowald and Tiefert, 1970), and immunological studies on the source of certain yolk proteins (Gelti-Douka *et al.*, 1974). These studies have shown that the oocyte nucleus is relatively inactive in RNA synthesis, the nurse cells being the major source of ribosomal and messenger RNA for the developing egg. The follicle cells play a role in facilitating the uptake of yolk proteins by the oocyte from the hemolymph, and may also synthesize proteins which are contributed to the developing egg (Anderson, 1971; Bast and Telfer, 1976). At the end of oogenesis, the follicle cells also secrete the protective coverings of the egg (King and Koch, 1963; Petri *et al.*, 1976). These aforementioned descriptions are necessarily somewhat general and are limited by the resolution of the particular techniques employed. The analyses of mutations affecting these processes provide an alternate approach in which specific functions can be identified by their absence in mutant individuals.

Among the many X-linked mutations isolated in the past years, one

which has a most dramatic effect on egg morphology is *fs(1)K10*
(Wieschaus, Marsh, and Gehring, in preparation). Eggs from normal
Drosophila females possess two protective coverings, an inner transparent
vitelline membrane, and an outer opaque chorion. The hexagonal pattern
seen on the chorion is an imprint left by the follicle cells which secrete
both it and the vitelline membrane (Fig. 2). At the anterior end of the egg
there are two dorsal appendages which probably serve as respiratory
organs (Hinton, 1969). Eggs laid by *K10* females lack these dorsal
appendages and instead have a huge mass of chorionic material secreted
at the anterior end where these appendages would normally be found.
The vitelline membrane is also abnormal. Rather than tapering
anteriorly, it ends in a bump. *K10* eggs are rarely fertilized and in the few
cases where this does occur, the embryo becomes disorganized at
gastrulation when supernumerary folds appear.

Since the most apparent abnormalities in *K10* eggs are found in the
chorion and the vitelline membrane, one might expect the *K10* mutation
to block a function associated with the follicle cells. This hypothesis can
be tested by making mosaic flies in which only the germline or the
somatic tissues are of mutant genotype. We constructed these mosaics by
transplantation of the embryonic precursors for the germline (pole cells)
between mutant and wildtype embryos, using techniques developed by
Illmensee (1973) and refined for our purposes by Van Deusen (1976).
These pole cells can be easily identified as a cap of slightly larger cells at

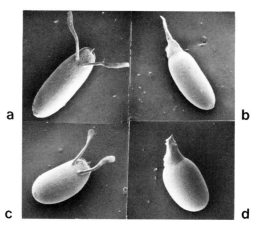

Fig. 2. Eggs from mutant and normal *Drosophila* females. A wild-type *Drosophila* egg possesses two
dorsal appendages at the anterior end of the chorion (a). In eggs laid by females homozygous for *fs(1)K10*,
these appendages are much enlarged and fused (b). Eggs laid by *1(1)SE* homozygotes are much shorter
and fatter than normal (c). Females homozygous for both mutations *(fs(1)K10* and *1(1)SE)* lay eggs of *K10*
morphology which are short and fat (d). The defects are therefore additive and the mutations affect
different non-interacting processes.

the posterior pole of the embryo at the blastoderm stage. Since donor and host strains carried different genetic markers, successful germline transfers could be identified by progeny testing each prospective host. In the first experiment wild-type pole cells were transplanted into mutant hosts. Of the 18 embryos homozygous for $K10$ which survived to the adult stage, nine proved to be mosaic and yielded eggs of normal as well as $K10$ phenotype (Table I, from Wieschaus, Marsh, and Gehring, in preparation). The normal eggs developed into adults carrying the donor marker genotype, proving that they were derived from transferred donor pole cells. These donor cells thus were able to make normal eggs even though they were surrounded by somatic cells which were homozygous for $K10$. In the reciprocal experiments where mutant pole cells were transferred to wild-type somatic hosts, $K10$ germ cells produced $K10$ eggs which gave no indication of any amelioration of the mutant phenotype by the surrounding wild-type somatic cells. In both experiments the morphology of the eggs produced by the transferred pole cells correlated strictly with their genotypes and was independent of the genotype of the somatic tissues of the fly. Thus, the altered morphology of the chorion and vitelline membrane results from a block in what is normally a germline function. These results suggest that the oocyte nurse cell complex provides cues which influence the pattern of secretion by the follicle cells which overly it. This hypothesis can be refined as more mutations become available which affect the morphology of the chorion. It might also be tested by pole cell transplants between closely related *Drosophila* species which show different chorion

TABLE I

K10 germline mosaics constructed by pole cell transplantation

Injected Embryos	Surviving Hosts	Mosaics
161	18 ♀♀	9 ♀♀ laid normal eggs as well as K10 eggs
262	36 ♀♀	4 ♀♀ laid K10 eggs as well as normal eggs

When wild-type pole cells carrying cuticle markers were transferred to embryos homozygous for $K10$, the resultant females with mosaic germlines produced normal as well as $K10$ eggs. The progeny which developed from these normal eggs all showed the donor cuticle markers, confirming that they were derived from transferred donor pole cells. Similarly $K10$ homozygous pole cells transferred to wild-type embryos produce eggs of $K10$ morphology. Both experiments indicate that the $K10$ morphology depends strictly on the genotype of the germ line and not on that of the somatic cells which secrete the chorion. The crosses which yield the homozygous $K10$ embryos also produce heterozygous female embryos and male embryos. While mosaics involving these types serve as useful controls for the efficiency of the transplantation, they are uninformative with respect to $K10$ and have been omitted from the table. Complete details of all transfers and cuticle markers used can be found Wieschaus, Marsh, and Gehring (in preparation).

morphologies. The Hawaiian sub-group might provide excellent opportunities for this type of work since differences in chorion morphologies have been characterized at the electron microscope level (Kambysellis, 1975), and transplantation of ovaries have shown that at least certain interspecific combinations are capable of vitellogenesis and normal egg production (Kambysellis, 1970).

Pole cell transplants have also been used to demonstrate the germline dependence of the maternal effects associated with the mutations *deep orange (dor)* and *maroonlike (mal)* (Marsh *et al.*, 1977; Marsh and Wieschaus, 1977). In addition to the eye color effect for which it is named, *dor* causes sterility in females homozygous for the mutation. *Dor* embryos from homozygous mothers become abnormal after gastrulation and die before hatching (Counce, 1956). The mutation *mal* does not affect the viability of the flies, but *mal+* product is necessary for the activity of aldehyde oxidase, xanthine dehydrogenase, and pyridoxal oxidase (Glassman and Mitchell, 1959; Courtright, 1967). Although the lack of xanthine dehydrogenase results in brownish eye color, normal oogenesis supplies enough *mal+* factor to the egg to ensure the normal eye color of the offspring. Brown eyed progeny are found only among the progeny of homozygous females who were unable to provide *mal+* factor to the eggs. Both *dor* and *mal* are rescuable by the paternal genome and *dor* can even be rescued by injection of cytoplasm from wild-type eggs (Garen and Gehring, 1972). Thus, the mutations presumably affect metabolic factors which must be put in the egg and thus differ from mutants like *K10* which affect the egg structure itself.

These three mutations do not provide us with a very broad sample of oogenetic functions and allow no generalizations about the role of the germline in oogenesis. Pole cell transplantations are probably too tedious for testing the large numbers of mutations necessary for an adequate survey. Even in the *K10* experiment where the survival and transfer success both ranged from 30% to 50%, it was necessary to inject 423 embryos to obtain the twelve successful cases on which the conclusions were based. There are several reasons for this. Since maternal effect mutations are usually maintained over balancer chromosomes in heterozygous stocks, the collections from which the donor or host is drawn contains a mixture of heterozygous and homozygous embryos. At the stage when the transplants are made, neither the genotype nor the sex of the embryo can be determined. Thus, a large fraction of the transfers are expected by chance to be uninformative combinations involving heterozygous embryos or cases where either the donor or host is male.

A more efficient way of producing mosaics in *Drosophila* involves the use of mitotic recombination (Stern, 1936), a technique which has been used extensively to study the development of imaginal discs. Its use in studying the role of the germline in oogenesis can best be described using two of the maternal effect mutants which we have tested earlier in pole cell transplants, fs(1)K10 and mal. Both these mutants are recessive and a heterozygote produces eggs of normal morphology and enzyme activity. If trans-heterozygous individuals are X-irradiated at some point during their development, a random recombinational event can occur in a single cell, such that the two daughter cells formed after mitosis are each homozygous for one homologue and the recessive maternal effect mutation it carries (Fig. 3). If this occurs in the germline, one of the daughter cells will produce K10 eggs, the other daughter, eggs of normal shape which give rise to brown eyed mal progeny. Recombination of course can produce clones in other tissues as well, but since K10 and mal have effects which depend strictly on the genotype of the germline, such clones will go undetected. Mitotic recombination is an extremely efficient way of producing mosaics. Large numbers of young larvae can be irradiated, and with a dosage of 1000 r, about 20% of the adults will have germline clones (Wieschaus and Szabad, in preparation).

In collaboration with Gans and coworkers, I have begun a survey of the tissue specificity of a large number of maternal effect mutations by asking whether germline clones for these mutations in heterozygous flies still produce their associated defects. Since mitotic recombination can also occur in follicle cells or other cell types involved in oogenesis, an independent marker is needed to ascertain whether or not the clone is in the germline. To this purpose we have coupled the particular mutant to be tested with K10. Germline clones arising in this cis-heterozygote will

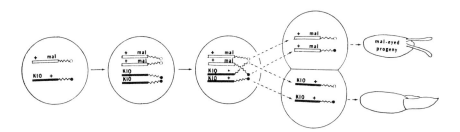

Fig. 3. Mitotic recombination in the female germline. A germ cell heterozygous for fs(1)K10 and mal will produce eggs of normal morphology and normal enzyme activity. Mitotic recombination can be induced in single cells if the individual is X-irradiated during its development. If the crossover occurs between mal and the centromere as indicated in the drawing, the recombination results in two daughter cells, one homozygous for K10, the other for mal. Females with such a clone in their germline will lay some eggs of K10 morphology and others which develop into brown eyed mal progeny (see text).

be homozygous for both mutants (Fig. 4) and any eggs produced by such clones will be *K10* in morphology. These eggs may also express the phenotype associated with the mutant to be tested, but only if the mutant affects a germline function. If the mutation depends on the genotype of the follicle cells, the egg will not show the associated defect since the follicle cells, like all somatic cells, have remained heterozygous.

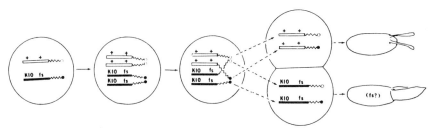

Fig. 4. Rationale for using mitotic recombination to test the tissue specificity of different female sterile mutations. Females heterozygous for *K10* and a female sterile mutation *(fs)* coupled in *cis* configuration will lay eggs of normal phenotype. Clones homozygous for *fs* will also be homozygous for *K10* and if mitotic recombination occurs in the germline, eggs formed from such clones will be *K10* in phenotype. If the *fs* mutation interferes with a germline function, the germline homozygosity will be sufficient to block the development of the clonally derived *K10* eggs or to cause them to show the particular pattern of defects associated with the mutation. If on the other hand the mutation blocks functions which normally occur in somatic cells, these functions will be unaffected by homozygosity in the germline and clonally derived *K10* eggs will be laid at normal frequency and will show no trace of the defects associated with the mutant tested.

Table II summarizes the results obtained from tests of three different maternal effect mutations. Females homozygous for the first mutant tested, *fs(1)1304 (=1304)* are blocked during the early stages of vitellogenesis and lay no eggs (Gans *et al.,* 1975). Histological examination of the ovaries from such females reveal many degenerating cysts in which the nurse cells have pycnotic nuclei and in later stages the follicle cells are disorganized (Khippel and King, 1976). Heterozygotes for the double mutant *K10 1304* lay perfectly normal eggs, indicating that *1304* like *K10* is totally recessive. The animals were irradiated as young larvae with a dosage of 1000 r. We know from controls containing *K10* but not *1304* that this dosage produces mitotic recombination resulting in about 3 *K10* eggs per 1000 eggs examined. Germline clones induced in *K10 1304* heterozygotes will be generally homozygous for both mutants. If the *1304* function is germline, oocytes derived from these clones will be blocked during early oogenesis and mosaic females will lay no *K10* eggs. The marked reduction of *K10* eggs obtained from irradiated heterozygotes indicates that this is the case, i.e., that *1304* causes egg retention in mosaics when only the germline is homozygous for the mutation and thus the gene function eliminated by the mutant is active in

TABLE II

Mitotic recombination used to test the germline specificity of female sterile and maternal effect mutations.

Mutation Tested	Phenotype of Female Sterile Mutation	Heterozygous Females Tested	Total Eggs Examined	Frequency of K10 Eggs per 1000
Control (K10 fs+)		176	15,047	3.2
K10.1304	blocked during oogenesis	151	9,814	0.4
K10 384	eggs flaccid, lack dorsal appendages	157	12,522	2.2 (eggs K10, but not flaccid)
K10 1(1)SE	eggs short and fat	281	20,522	4.0 (eggs K10, but not short)

Heterozygous females *(K10 fs/X) were irradiated with 1000 r as second instar larvae. Germline mosaics were detected as adult females which laid K10 eggs during the 2-4 day test period. Since most clones homozygous for K10* should also be homozygous for the *fs* mutation to be tested, the frequency and morphology of the clonally derived eggs in the experimental series indicate the effect of germline homozygosity for the mutant in mosaics where the somatic tissue has remained heterozygous. The marked reduction in clonally derived K10 eggs when K10 was coupled with 1304 implies that the block in oogenesis found in 1304 homozygotes is germline dependent. Conversely, the failure to detect the defects associated with 384 and 1(1)SE in tests involving those mutants implies that these defects are produced only when some tissue other than the germline is homozygous. Thus 1304 seems to block a germline function while 384 and 1(1)SE are involved in the somatic cell contribution to oogensis.

the oocyte nurse cell complex.

The germline dependence of *1304* contrasts with the results obtained for the second and third mutants listed in Table II. Females homozygous for *fs(1)384 (=384)* lay eggs which are flaccid and lack dorsal appendages (Gans *et al.,* 1975). Eggs from the double homozygotes K10 384 are also flaccid and show only an extremely rudimentary K10-type dorsal appendage. The defects characteristic of each mutant can be dissociated from each other in mosaics in which the germline is homozygous for both. The eggs which develop from such clones are K10 in phenotype but show none of the defects associated with 384. The eggs are not flaccid and the dorsal appendage material, although K10 in phenotype, is of normal size. The defects associated with 384 are independent of the genotype of the germline and thus the locus codes for functions performed elsewhere by the somatic tissue of the fly. A similar conclusion is drawn for the mutant *1(1)SE,* which at 22° results in unusually short, fat eggs. Eggs which develop from germline clones of this mutant are of normal length indicating that the factors which reduce the size of the mutant egg depend on cells of somatic origin.

III. PROSPECTS, PROBLEMS, AND OTHER MOSAIC SYSTEMS

As more mutants are tested it should be possible to determine what functions or more accurately what classes of mutant defects are follicle cell dependent and which depend on the germline. The mutations tested so far were recognized as affecting oogenesis because homozygous females produced either no eggs or only abnormal eggs or progeny. This test requires that the homozygotes survive to the adult stage and consequently, at least under our conditions, the mutations can have no major effect on the viability of the mother. There are however many other genetic loci involved in oogenesis which affect earlier developmental processes as well. Mutations in these loci might result in death before the adult stages when their effect on oogenesis can be assessed. When temperature sensitive alleles are available, the females can be raised at permissive temperatures and the temperature shift made only with adults to study the effect of the mutation on oogenesis. Mosaics provide an alternate method to study the role of these loci since females with small patches of homozygous tissue in their germline would be expected to survive. Such an analysis should also be possible in the follicle cells, and other cell types involved in oogenesis, once marker mutations analogous to K10 are found which allow identification of clones in those tissues.

One of the disadvantages of mitotic recombination compared to pole cell transplants is that it is only possible to make relatively small patches of mutant tissue in a wild-type (heterozygous) fly. Mitotic recombination generally does not provide a technique for transferring a single wild-type cell into an entirely mutant background, as pole cell transplantations do. The analyses we have just described are based on the assumption that oogenetic functions are either germline or somatic or at least that a particular defect associated with a maternal effect mutation results from malfunction in only one of these cell types. The experiments give ambiguous results when a mutation causes abnormal oogenesis if either tissue is mutant or when a plus allele in either cell type can ensure normal development of the egg. The possibility of such dual effects can be eliminated, as it was in the initial experiments with K10, by performing reciprocal pole cell transplants.

In *Drosophila* it is also possible to make mosaics by transplanting the larval or adult gonad between individuals mutant and wild-type for different maternal effect mutations (Clancy and Beadle, 1937; Clancy and Welborn, 1948; King and Bodenstein, 1965; Smith *et al.*, 1965; Klug *et al.*, 1968; Garen and Gehring, 1972; Swanson and Poodry, 1976). Since the gonad in these stages contains the precursors for both the follicle cells and germline components, the transplants do not distinguish specifically between germline and soma. They are useful in distinguishing the contributions of different somatic tissues since mutant ovaries in a wild-type environment should not produce defective eggs if the mutation blocks processes which occur in somatic cells outside the ovary. Examples of such cell types might be the fat body which is thought to produce the yolk proteins carried by the hemolymph into the egg and the corpus allatum, the source of juvenile hormone necessary for oogenesis. Most mutants behave autonomously in gonad transplants and thus affect functions of cells in the ovary itself, be they somatic or germline. However, one mutant *(apterous-4)* (King and Bodenstein, 1965) has been found to depend on tissue outside the ovary and is thought to be involved in juvenile hormone metabolism.

Mosaics can also be made in *Drosophila* by loss of chromosomes during early cleavage. The gynandromorphs used in most of these studies arise from the loss of an X-chromosome and are thus mixtures of male and female cells. They are therefore not very suitable for an analysis of oogenesis. Totally female mosaics can be constructed by loss of a Y chromosome (Baker, 1975) since both XX and XXY individuals are fertile females in *Drosophila*. The Y normally does not carry genes necessary for female fertility and such an analysis necessitates genetic translocations where the particular locus to be studied has been transferred to the Y. An alternate procedure particularly useful for X-linked markers would be an X-chromosome loss in triplo-X females (meta-females) making mosaics of XX and XXX tissue. Although XXX individuals are poorly viable, XXX tissue survives in mosaics and XXX germ cells make approximately normal numbers of eggs (Schupbach *et al.,* in preparation). The loss of the X in this system might be used to uncover recessive mutations homozygous on the remaining two X chromosomes. The principal advantage of the chromosome elimination method of producing mosaics is that since the elimination occurs early, the regions of mutant and wild-type tissue are roughly the same size. The final mosaic pattern records chance events in the establishment of the individual primordia which are random with respect to genotype of maternal effect markers. This allows a practically unlimited spectrum of mosaic females in which, for example, the entire follicle cell population or the entire fat body might be of mutant genotype.

Since most of the RNA synthesis in the germline during oogenesis occurs in the nurse cell nuclei, one expects most mutations which test as germline dependent to block nurse cell functions. This expectation can potentially be tested in mosaic cysts where the nurse cells and oocytes differ in genotype. Such mosaics cannot be constructed mechanically by transplantation given that the entire complex arises mitotically from a single cell and the 16 sister cells remain interconnected by cytoplasmic bridges. Mosaic cysts might be produced using mitotic recombination during egg production in the adult. However, since different sectors of the nurse cell complex are generated with each of the four consecutive mitoses, it is impossible to obtain from a single mitotic recombination a mosaic cyst in which the nurse cells are entirely of mutant genotype.

By combining the different type of mosaics described above, it should be possible in the future to determine with considerable accuracy the tissue in which a particular maternal effect mutation acts. Fig. 5 summarizes the different cell types involved in oogenesis and the mosaic techniques available to distinguish between them.

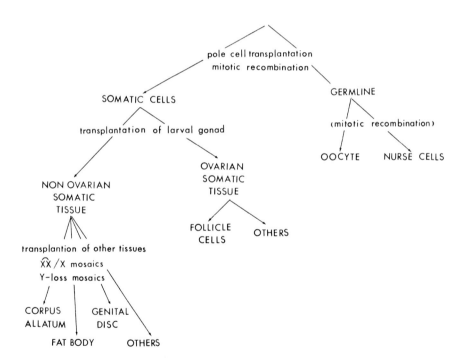

Fig. 5. Mosaic techniques for distinguishing the tissue specificity of maternal effect mutations. The different cell types are printed in large lettering. Their branching arrangement does not necessarily imply cell lineage relationships during development although they often correspond to such relationships. Techniques which can be used to distinguish between two cell types are superimposed on the branch which divide them. Potential useful techniques which have not yet been applied to tissue specificity in oogenesis are printed in parentheses.

IV. DEVELOPMENT AND FUNCTION OF
THE FEMALE GERMLINE

In the preceding section we have described the use of genetic mosaics to identify the primary site of action of certain maternal effect mutations. Most of these mutations blocked functions which were probably specific to the adult stage and limited to oogenesis as such. However, certain aspects of the egg structure, such as the axis of polarity, are probably influenced by the structure of the ovary and by processes and genes active at early stages during the development of the gonad. One of the advantages of X-ray induced mitotic recombination is that the mosaics can be produced at any time during development when the cells are mitotically active. It can thus be used not only to define tissue specificity of particular mutations, but by comparing the patterns of clones induced at different stages, to draw conclusions about intervening developmental events.

When the pole cells are first budded off at the posterior of the embryo, they number about ten (Rabinowitz, 1941; Sonnenblick, 1950; Poulson, 1950; Turner and Mahowald, 1976; Zalokar and Erk, 1976). During the next hour, they continue to divide, albeit at a slower rate then their somatic counterparts. When the blastoderm stage is reached they number about 40. During gastrulation the pole cells undergo a complicated migration into the interior of the embryo, where they are joined by somatic cells of mesodermal origin to form the gonad. Not all the cells seen at the surface complete the migration into the gonad. The remainder are thought to be lost or to contribute to the cuprophilic cells of the midgut (Poulson, 1950).

We have used mitotic recombination to determine the number of pole cells which contribute to the functional germ line in the adult, and whether a cell which does contribute to the germline is likely to contribute to other structures as well (Wieschaus and Szabad, in preparation). Embryos heterozygous for *K10* and *mal* were irradiated at the blastoderm stage. Among the 544 adult females whose egg production we examined, eighteen were found which yielded some *K10* eggs and were therefore determined to be germline mosaics. The fraction of the germline homozygous for *K10* could be estimated from the fraction of the eggs laid by each female which were *K10* in phenotype. If all of the 40 pole cells observed at the surface contribute to the functional germline, we would expect the *K10* eggs laid by each mosaic female to represent only 1/80 of her total egg production since the *K10* eggs are only half the progeny of the irradiated pole cell. However, the average clone sizes were much larger (Fig. 6) and indicated instead that about eight pole cells contribute to the germline. Because *K10* is located on the distal most tip of the X-chromosome, virtually all recombinational events will result in homozygous *K10* clones, regardless of where along the chromosome they occur. The mosaic females were also originally heterozygous for *mal* and a certain fraction of *K10* clones should be associated with a homozygous *mal* twin clone. From work on imaginal discs it is known that about 25% of the mitotic recombinations on the X chromosome occur between the *K10* and *mal* loci and thus only 75% of these events would result in clones homozygous for *mal* (Garcia-Bellido, 1972; Becker, in press). If both daughter cells of an irradiated cell remain in the germline and produce eggs in the adult, about 75% of the mosaic females which produce *K10* eggs should give rise to brown eyed *mal* progeny. Seventy five percent is very close to the figures obtained when

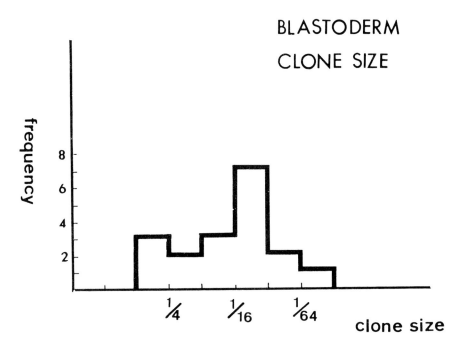

Fig. 6. Size of clones induced at the blastoderm stage in females heterozygous for *K10*. The fraction of the germline homozygous for *K10* was estimated from the fraction of the eggs produced by mosaic females which were *K10* in phenotype. These clone sizes are inversely proportional to the number of prospective germ cells at the time of the irradiation. Since the clones whose sizes are depicted here represent only half the progeny of the irradiated blastoderm cell, the average clone size must be divided by two before estimating the number of pole cells which contribute to the germline. The data are based on examination of 588 females and do not include three small clones of one *K10* egg each which evidently arose spontaneously during later development. A similar frequency of spontaneous clones (2/284) was also found in unirradiated controls.

irradiations were made at the blastoderm stage or later during embryonic or early larval development (Wieschaus and Szabad, in preparation). This means that although less than a quarter of the pole cells observed at the blastoderm surface contribute to the germline, those that do, contribute both daughters. This makes it rather unlikely that cells which contribute to the germline contribute their progeny to other structures as well.

The frequency with which germline clones were produced in individuals irradiated at the blastoderm stage (4%) was much lower than the frequency obtained after larval irradiation (18%). Since the frequency of clonal induction is directly related to the number of target cells, the higher frequencies found with larval irradiations suggest that the cells have begun to divide. This proliferation can also be followed by decrease in clone size since the larger the number of germ cells, the smaller the relative contributions of any individual cell. From our cell lineage data we estimate that 48 hours are required during larval development for the cell number to double. This rate of increase continues up to the onset of pupation (Fig. 7). At pupation, when there are about 100 cells, the number ceases to increase and the same number of germ cells are found in adults. This is not due to a cessation of mitosis, but to a switch from the logarithmic growth divisions of the larva to that more characteristic of adult stem cells. In the embryo and larva both daughter cells remain in the dividing population and the cell number increases by a factor of two with each round of division. After a stem cell division we expect only one of the two daughter cells to remain in the dividing stem cell population. The other begins its differentiation into an oocyte. The differential nature of the stem cell division is most clearly seen in the pattern of *K10* eggs obtained from mosaics induced when heterozygous adults are irradiated. Since the daughter cell which forms the oocyte should be chosen randomly with respect to genotype, 50% of the mitotic recombinations should result in single *K10* eggs, and 50% in a series of *K10* eggs laid over the remainder of the female's life. The distributions obtained after adult irradiations are in fact bimodal with peaks at 1 *K10* egg and 5-8 *K10* eggs (Fig. 8). Since the frequency of clonal induction is high after irradiation of the adult, many females would be expected to possess two independent clones and thus the frequency of females laying single *K10* eggs is less than 50%. However when the Poisson distribution is used to predict the number of mosaic females laying 1, 2, and 3 or more *K10* eggs, the results obtained fit very well with those predicted from a one to one ratio of single *K10* eggs to *K10* egg series.

Cell lineage techniques provide a unique opportunity to define the stem cell division, and to follow the transition between the logarithmic and stem cell phases of germline division. The stem cell divisions would

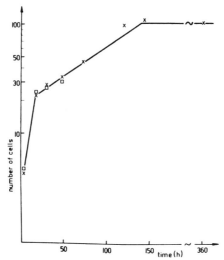

Fig. 7. The logarithmic increase in the number of germ cells during larval development. The cell number at each stage was calculated as the reciprocal of arithmatic average of clone size. At the onset of pupation, the number becomes constant at 105 cells. Data based on examination of 2006 females irradiated at different embryonic, larval, pupal, and adult periods.

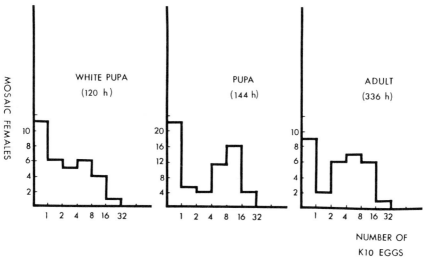

Fig. 8. Number of K10 eggs laid by individual mosaic females irradiated as prepupae, 24 hour old pupae, and adults. The mosaics induced during pupal and adult periods fall into two classes;, those laying single K10 eggs and those laying between 5 and 16 K10 eggs. This bimodal distribution is characteristic of stem cell divisions. Although many mosaic females are found to lay single K10 eggs after prepupal irradiations, the clone size distribution is not yet clearly bimodal. Instead it probably reflects the transition into the stem cell type divisions from the growth divisions which occur during larval development. Data from 205, 258, and 130 females irradiated as white pupae, pupae, and adults.

seem to begin in the ovary a little after the onset of pupation, judging from the observation that the cell numbers do not increase after that stage. By 24 h after pupation clone sizes are clearly bimodal (Fig. 8) and show a good fit with the distribution predicted from the Poisson. The first signs of cytoplasmic bridges between oocytes and nurse cells are observed at about this time (King, 1970). The distributions during embryonic and larval development by contrast show only a single modal clone size, and only at the prepupal stage are a substantial number of single K10 egg clones found (Fig. 8). Although the distribution of prepupal clone sizes is not yet clearly bimodal, the presence of these single egg clones may indicate that some of the cells have begun to divide in stem cell fashion or that a fraction of the irradiated germ cells are sensitive to mitotic recombination but do not divide until the stem cell divisions actually begin. It is also possible that not all the germ cells in the pupal gonad are destined to remain in the ovarioles as active stem cells in the adult. When the ovarioles are formed, the germ cells which do not join the stem cell population would make eggs immediately. Clones induced in these cells would form only one K10 egg. The possibility that some pupal germ cells do not become stem cells was suggested by King (1970) from histological observations. If true, it would imply that normal eggs can be produced without stem cell divisions and that the polarized nature of that division is not causally related to the polarized axis of the egg. The data presently available does not allow us to describe unambiguously the manner in which the transition into the stem cell mode of division is made or to choose between these models. Indeed it is even possible to develop alternative, more complicated models to explain the bimodality and constant cell numbers found in the late pupal and adult stages. A more detailed analysis of the pupal period would be informative, particularly one using markers which would allow one to follow the fate of the non-K10 daughter cell. Mal is probably not ideal for such studies because the maternal effect is not 100% penetrant and small clones homozygous for mal cannot be reliably detected.

From the clone size obtained after adult irradiations we calculate that each adult female possesses about 100 germline stem cells, or about 50 per ovary. Each ovary is seen in histological preparation to be made up of 17 ovarioles, each with its own germarium and its own complement of stem cells. From the difference between the total number of ovarioles and the total number of germ cells, we can conclude that most of the ovarioles contain two or three stem cells. Support for this conclusion was obtained when germline clones were identified in situ using the lack of aldehyde oxidase activity associated with the mutation mal. Ovarioles which contained non-staining mal cysts invariably contained staining *mal+* cysts as well, regardless of whether the clones were induced in the adult or in late larvae which were then allowed to develop to adults before staining (Fig. 9). No ovarioles were found after irradiation in these stages which contained only *mal* cysts as would be expected if an ovariole possessed a single stem cell in which the recombination occurred. The developing oocyte-cysts in an ovariole are arranged according to age with the most advanced at the basal end, and the pattern of mosaicism among the cysts presents a record of the order at which the differently marked stem cells in each ovariole divide. If the two or three stem cells alternate in their divisions, we expect certain specific repeated patterns, either alternating blue (*mal+*) and white (*mal*) cysts or two blue cysts always followed by a white one. These patterns are not found. Instead it would appear from a large analysis of mosaic ovarioles in Triplo-X Diplo-X flies (Schupbach, *et al.,* in preparation) that each cell tends to divide in bursts of activity, followed by periods of quiesence when the other stem cells divide. This is seen by the high frequency with which neighboring cysts are of the same genotype.

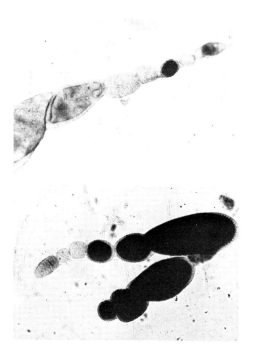

Fig. 9. Mosaic ovarioles produced by irradiation of larvae heterozygous for *mal*. Homozygous *mal* oocyte cysts were detected by their lack of aldehyde oxidase activity. Mosaic ovarioles could only be produced if each ovariole contains more than one germline stem cell.

V. CONCLUDING REMARKS

The work described in this chapter has concentrated principally on the female germline. In the first section we have shown how mutations can be used as probes to define different processes in oogenesis. Then using pole cell transplantation or mitotic recombination, we constructed mosaic flies in which either the germline or the somatic tissues were of mutant genotype. This allowed us to determine which tissue was responsible for the defect and by extrapolation the tissue in which the normal function occurs. Although the tissue in which the mutation acts can usually be correlated with the type of defect observed in the ovary or in the egg, certain mutations gave unexpected results. For example, *fs(1)K10* proved to be germline dependent, in spite of its obvious effects on the chorion and vitelline membrane. Since other aspects of chorion morphology are controlled by somatic tissues (i.e., the defects associated with fs(1)384), the construction of the chorion involves an interaction between the two cell types. This may well be true for most processes in oogenesis. The isolation of maternal effect mutations and their use in mosaics provide a very refined technique for dissociating different aspects of these interactions.

The characterization of mutations like *K10* and *maroonlike* whose effects depend strictly on the genotype of the germline, has provided us with the marker mutations necessary for the cell lineage analysis of germline development, which forms the second part of this chapter. By using mitotic recombination to mark single cells at the blastoderm stage, we have shown that the female germline arises from only a small fraction of the pole cells observed at the surface of the embryo. During larval development this number increases logarithmically since after each division both daughter cells remain in the undifferentiated dividing population. The situation changes at pupation, where the number of cells becomes constant and the clone sizes begin to assume the bimodal distribution characteristic of stem cell divisions. From the total number of stem cells in the adult, we have calculated that each ovariole contains 2 or 3 stem cells, a conclusion supported by the production of mosaic ovarioles where the clonally derived cysts were identified histochemically.

ACKNOWLEDGMENTS

I would like to thank my collaborators Larry Marsh, Trudi Schupbach, and Janos Szabad for their specific contributions to this paper, and Jack Levy and Lynn Littlefield for helpful comments on the manuscript. My own research is supported by a Swiss National Science Foundation Grant to Rolf Nothiger (No. 3. 741. 76). During the writing of this manuscript I was supported by PHS Training Grant No. HD 07029 from the National Institutes of Health, HEW.

REFERENCES

Anderson, L. M. (1971). *J. Cell Sci.* **8**, 735-750.
Baker, B. S. (1975). *Genetics* **80**, 267-296.
Bakken, A. H. (1973). *Develop. Biol.* **33**, 100-122.
Bast, R. E. and Telfer, W. H. (1976). *Develop. Biol.* **52**, 83-97.
Becker, H. J. In: "The Genetics and Biology of *Drosophila*" (M. Ashburner and E. Novitski, eds.). Academic Press, New York (in press).
Bier, L. (1963). *Wilhelm Roux' Arch.* **154**, 552-575.
Bucher, N. (1957). *Rev. Suisse Zool.* **64**, 91-188.
Clancy, C. W. and Welborn, W. S. (1948). *Genetics* **33**, 606.
Clancy, C. W. and Beadle, G. W. (1937). *Biol. Bull.* **72**, 47-56.
Counce, S. J. (1956). *Z. Indukt. AbstVererbungsl.* **87**, 443-461.
Courtright, J. B. (1967). *Genetics* **57**, 25-39.
Gans, M., Audit, C. and Masson, M. (1975). *Genetics* **81**, 683-704.
Garcia-Bellido, A. (1972). *Molec. Gen. Genet.* **115**, 54-72.
Garen, A. and Gehring, W. (1972). *Proc. Nat. Acad. Sci. U.S.* **69**, 2982-2985.
Gelti-Douka, H., Gingeras, T. R. and Kambysellis, M. P. (1974). *J. Exp. Zool.* **187**, 167-172.
Glassman, E. and Mitchell, H. K. (1959). *Genetics* **44**, 153-162.
Hinton, H. E. (1969). *Ann. Rev. Entomol.* **14**, 343-368.
Illmensee, K. (1973). *Wilhelm Roux' Arch.* **171**, 331-343.
Janning, W. (1976). *Wilhelm Roux' Arch.* **179**, 349-372.
Kambysellis, M. P. (1970). *J. Exp. Zool.* **175**, 169-180.
Kambysellis, M. P. (1975).
Kaplan, W. D., Seecof, R. L., Trout III, W. E., and Pasternack, M. E. (1970). *Am. Nat.* **104**, 261-271.
Khipple, P. and King, R. C. (1976). *Int. J. Insect Morphol. Embryol.* **5**, 127-135.
King, R. C. (1970). "Ovarian Development in *Drosophila melanogaster.*" Academic Press, New York.
King, R. C. and Bodenstein, D. (1965). *Z. Naturforsch.* **20b**, 292-297.
King, R. C. and Burnett, R. G. (1959). *Science* **129**, 1674-1675.
King, R. C. and Koch, E. A. (1963). *Quart. J. Micros. Sci.* **104**, 297-320.
King, R. C. and Mohler, J. D. (1975). In: "Handbook of Genetics" (R. C. King, ed.). Vol. 3 Plenum Press, New York.

Klug, W. S., Bodenstein, D., and King, R. C. (1968). *J. Exp. Zool.* **167**, 151-156.

Koch, E. A. and King, R. C. (1966). *J. Morphol.* **119**, 283-303.

Mahowald, A. (1972). *J. Morphol.* **137**, 29-48.

Mahowald, A. and Tiefert, M. (1970). *Wilhelm Roux' Arch.* **165**, 8-25.

Marsh, J. L. and Wieschaus, E. (1977). *Develop. Biol.* **60**, 396-403.

Marsh, J. L., van Deusen, E. G., Wieschaus, E., and Gehring, W. J. (1977). *Develop. Biol.* **56**, 195-199.

Mohler, J. D. (1977). *Genetics* **85**, 259-272.

Petri, W. H., Wyman, A. R., and Kafatos, F. C. (1976). *Develop. Biol.* **49**, 185-199.

Poulson, D. F. (1950). In: "Biology of *Drosophila*" (M. Demerec, ed.), pp. 168-274. Wiley, New York.

Rabinowitz, M. (1941). *J. Morphol.* **69**, 1-49.

Rice, T. B. (1973). Ph.D. Thesis. Yale University, New Haven, Connecticut.

Schupbach, T., Wieschaus, E. and Nothiger, R. (in preparation).

Smith, P. A., Bodenstein, D. and King, R. C. (1965). *J. Exp. Zool.* **159**, 333-336.

Sonnenblick, B. P. (1950). In: "Biology of *Drosophila*" (M. Demerec, ed.), pp. 62-167. Wiley, New York.

Stern, C. (1936). *Genetics* **21**, 625-730.

Swanson, M. M. and Poodry, C. A. (1976). *Develop. Biol.* **48**, 205-211.

Turner, F. R. and Mahowald, A. P. (1976). *Develop. Biol.* **50**, 95-108.

Wieschaus, E., Marsh, J. L., and Gehring, W. J. (in preparation).

Wieschaus, E. and Szabad, J. (in preparation).

Van Deusen, E. B. (1976). *J. Embryol. Exp. Morphol.* **37**, 173-185.

Zalokar, M. (1960). *Exp. Cell Res.* **19**, 184-186.

Zalokar, M. and Erk, I. (1976). *J. Microsc. Biol. Cell* **25**, 97-106.

Cell Lineage and Homeotic Mutants in the Development of Imaginal Discs of *Drosophila*

Gines Morata[1] and Peter A. Lawrence

MRC Laboratory of Molecular Biology
Hills Road, Cambridge
CB2 2QH, England

I. INTRODUCTION

The larvae of *Drosophila* contain "imaginal discs", the cells of which secrete the cuticle of the cephalic, thoracic and genital segments of the adult fly. Each of them forms a specific and precisely defined piece of the adult cuticle; thus the imaginal discs can be named according to the fate of their cells: for example, the "wing disc" constructs the wing, notum and pleuras, and the "genital disc" the analia and genitalia. The develop-

---

[1]Present address: Centro de Biologia Molecular, C.S.I.C., Facultad de Ciencias, C — X, Universidad Autónoma de Madrid, Madrid 34, Spain.

ment of imaginal discs has been studied intensely in recent years by means of genetic mosaics. In particular, García-Bellido *et al.* (1973, 1976) discovered a new process that was called "compartmentalization", which in essence consists of the definition of groups of cells that become segregated from each other and at the same time determined to secrete specific parts of the cuticle. These specific regions are called "compartments". Each compartment begins existence as a small group of founder cells which generate a "polyclone" (Crick and Lawrence, 1975) of cells. A compartment is constructed by an entire polyclone. It was also shown (García-Bellido *et al.*, 1973, 1976) that a polyclone can become subdivided into two polyclones, a process which may be reiterated several times. As development proceeds, the wing disc cells become progressively subdivided into at least eight polyclones which form eight different compartments. Compartments have also been found in the leg disc (Steiner, 1976), the haltere disc (Morata and García-Bellido, 1976), the genital disc (Dubendorfer, 1977), and the labial disc (Struhl, 1977).

One important aspect of the compartment hypothesis is the relationship between compartments and homeotic genes (García-Bellido *et al.*, 1973, 1976). Mutations in those genes lead to substitutions of certain cuticular structures by others normally present elsewhere. The interpretation commonly accepted is that the homeotic genes control the acquisition of specific developmental routes by groups of cells (Lewis, 1964; Gehring and Nothiger, 1973). The failure, in mutant flies, of normal function results in the acquisition of an alternative route.

During the past three years we have been investigating the early development of compartments, and also the relationship between compartments and homeotic genes. Here we summarize the techniques used, state our main results, and present our ideas about the role of homeotic genes in the definition, segregation, and development of compartments.

II. TECHNIQUES

Three major techniques have been used in the analysis of wing disc development by means of genetic mosaics.

(1) *Gynandromorph analysis* is based on the loss of one X-chromosome in the first cleavage division of presumptive XX females. This loss produces XO tissue that is phenotypically male and uncovers any recessive marker mutant in the remaining X-chromosome. These marker mutants change the integument, making for example *yellow (y)*

bristles and cuticle instead of the dark brown colour, or *forked (f)* bristles and hairs instead of the straight wildtype. Thus male cells can be recognised almost everywhere in the adult cuticle. The most common method of generating *gynandromorphs* is to use a ring-shaped X-chromosome which is unstable and becomes lost early in development. Each landmark is male in one half of the gynandromorphs, and female in the other (Ripoll, 1972), showing that the mosaics are on the average 50% male and 50% female, and indicating that the ring is lost in the first cleavage division (Ripoll, 1972; Hotta and Benzer, 1972). However, as the orientation of the first cleavage division is random the likelihood that a given pair of landmarks will be of the same sex decreases as the distance between their progenitor cells (at the blastoderm) increases (García-Bellido and Merriam, 1969). Hotta and Benzer (1972) have proposed the term "sturt" as the unit for gynandromorph mapping. When a given pair of landmarks are of different sex in 1% of the gynandromorphs, their progenitor cells are separated by a distance of one sturt.

This method, invented by Sturtevant (1929), has been used to make blastodermal fate maps of cuticular structures (García-Bellido and Merriam, 1969; Ripoll, 1972; Hotta and Benzer, 1972; Wieschaus and Gehring, 1976a), and also with enzyme markers, to map internal organs (Janning, 1974a,b, 1976; Kankel and Hall, 1976). An extension has been developed (Hotta and Benzer, 1972) to locate the primary site of action (focus) of physiological or behavioral mutants. Gynandromorphs not only give information about position, but allow estimation of the number of progenitor cells in different organs; the more cells contribute to a given structure, the higher the probability that the structure will include both male and female cells. This value (the *frequency of mosaicism*) is taken as a measure of the original number of cells at blastoderm (Hotta and Benzer, 1972).

(2) *Clonal analysis* was first invented by Stern (1936) and developed by Becker (1957) for the study of cell lineage during development. It consists of marking single cells at different moments during development, the cell's descendants forming a clone that is recognised in the adult cuticle by the marker mutant phenotype. The most common method of generating clones takes advantage of X-rays which produce chromatid interchange between homologous chromosomes in the G2 period (*mitotic recombination*). By this method a cell heterozygous for a recessive marker mutant (and therefore phenotypically wildtype) produces, in the division following irradiation, two daughter cells, one homozygous for the marker mutant and the other for the wildtype allele (Fig. 1). The use of X-rays to produce marked clones has the

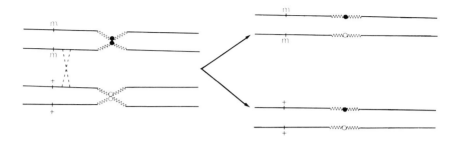

G2 stage of the mother cell   G1 stage of the two daughter cells

Fig. 1. *Mitotic recombination:* In G2 of the cell cycle when each homologous chromosome consists of a pair of chromatids, breakage and exchange occurs. Following mitosis and segregation of centromeres, one of the daughter cells can become homozygous for the mutant allele.

advantage that the event of clone initiation can be precisely timed. Analysing the frequency and size of clones produces a picture of growth and development of the imaginal discs (Bryant and Schneiderman, 1969; García-Bellido and Merriam, 1971, a,b), a technique which is equally applicable to analysis of abnormal discs like those of the homeotic mutants (Postlethwait and Schneiderman, 1971; Morata, 1975). This technique also gives information about cell determination in development; if two structures are included in a clone, the progenitor cell was clearly not exclusively determined to differentiate either structure (Gehring and Nothiger, 1973). However if clones induced at a later time never include both structures, this may indicate a determinative event, although here it has to be shown that clones are large enough and that there are no physical barriers between the two structures (Gehring and Nothiger, 1973; Morata and Lawrence, 1977a).

Clonal analysis can be extended by coupling the cell marker mutants with other mutants that affect development. This method can be used to study zygotic lethals (80% of zygotic lethals are viable in somatic spots, (Ripoll and García-Bellido, 1974)). It can also be applied to mutants, such as homeotic ones, that affect morphogenesis. We can, for example, study the behaviour of mutant clones for bx^3 (which transforms the anterior haltere into anterior wing) when surrounded by wildtype haltere cells, and analyse the cell autonomy of the mutant phenotype and time the requirement for the wildtype gene (Morata and García-Bellido, 1976).

(3) *The Minute technique* (Morata and Ripoll, 1975) is a variant of clonal analysis. We consider it separately because the information it gives is of a different nature. It makes use of dominant mutants called

Minutes of which there are about 50 different *loci* in *Drosophila*. They are recessive lethals, but in heterozygous condition have a phenotype of fine short bristles and delayed development. Most, if not all, of the delay occurs in the larval period, is different in each *Minute,* and increases the larval period between 20 and 80% (Ferrus, 1975). The division rate of *Minute* cells is reduced in proportion. It was found (Morata and Ripoll, 1975) that the normal division rate can be restored in wildtype clones (*M+*) even when they grow in *Minute* individuals. These *M+* clones compete out the surrounding *Minute* cells and reach a very large size. The *M+* clones are thus given the maximum opportunity to grow, so that the method tests the potential of individual cells over specific periods of development.

III. THE DEVELOPMENT OF THE THORACIC CUTICLE

A. *Embryonic Development*

We are concerned mainly with the mesothorax or second thoracic segment, and to a lesser extent with the metathorax or third thoracic segment. The mesothoracic cuticle is made by two imaginal discs, the wing and second leg discs. The metathoracic cuticle is also made by two discs, the haltere and third leg discs. A detailed description of these structures can be found in Ferris (1950).

Using gynandromorphs, Wieschaus and Gehring (1976a) showed that the progenitor cells of the different thoracic discs lie very close in the blastoderm. Moreover, the size of the disc primordia, measured by the frequency of mosaicism, is large and sturt distances within a given disc are sometimes larger than those measured between adjacent discs. These results suggest that at least some discs might share a common pool of cells at the blastoderm. This was demonstrated with clones produced during the blastoderm period (3h) (Wieschaus and Gehring, 1976b; Steiner, 1976; Lawrence and Morata, 1977). Marked clones sometimes extended from wing cuticle to second leg cuticle (both part of the mesothoracic segment), indicating that the mother cell was not specifically determined for either second leg or wing. However, no clone was found to extend between adjacent segments. All clones produced at 7h of development were restricted to the derivatives of single discs. These results suggest that the segments are determined at the blastoderm or a short time thereafter, whereas the discs appear later. However, we should emphasize a limitation; the marker mutants used in these experiments label only the adult cuticular structures, so it is not known whether the clones extend from the adult cuticle to the larval cuticle, or the internal tissues.

Minute+ clones produced at 3h or 4h of development can extend from anterior wing to anterior leg, or from posterior wing to posterior leg; however, no clone crossed from anterior wing to posterior wing, or from anterior leg to posterior leg. It follows that the anterior and posterior compartments must be defined in the thoracic segments before the segregation of wing and second leg takes place (Steiner, 1976; Lawrence and Morata, 1977). Probably, antero-posterior subdivision segregates two mesothoracic polyclones, part of whose derivatives are the adult cuticle, but which also include internal and larval tissues. This view is supported by the phenotype of homeotic mutants such as *bithorax* which transforms the anterior metathoracic compartment into the anterior mesothoracic one (Lewis, 1964); the transformation is not limited to the adult cuticle, but extends to larval tissues (Lewis, 1963, 1964) and to connections within the central nervous system (Palka and Lawrence, unpublished).

Thus there must be at least two segregations between the appearance of anterior and posterior polyclones at the blastoderm, or a short time thereafter, and the appearance of wing and second leg discs: by one, the presumptive adult disc cells become apart from the rest, the other segregates the dorsal cuticular cells (wing disc) from ventral cuticular cells (second leg disc). This order is suggested by the observation of Anderson (1963), in *Dacus* (another dipteran), that in early development there is only one adult primordium in the mesothorax that later becomes split into leg and wing discs.

Clearly the thoracic discs are formed by the cooperation of two groups of cells that are differently determined (there is never such a thing as simply "wing" or "leg" determination). This idea has led to the hypothesis ("construction hypothesis", Gehring, 1976) that the antero-posterior compartment boundary might be formed by the apposition in the adult of two groups of cells that were apart in the embryo and larva, in much the same way that left and right disc derivatives form a straight boundary where they meet in the midline of the fly. There are several arguments against this view; (1) The distance between adjacent anterior and posterior cells in the wing is about 7 sturts (García-Bellido and Ferrus, 1976; Lawrence and Morata, 1977), which is very close to the distance measured for the nearest landmarks of wing and second leg (6.2 sturts, Wieschaus and Gehring, 1976a). In this latter case we know that wing and leg cells overlap at blastoderm (Wieschaus and Gehring, 1976b). (2) The frequency of separation in gynandromorphs (the number of times in which a given structure is entirely of one sex and the other is entirely of the other sex) is a measure of the distance between the primordia (Wieschaus and Gehring, 1976a; Lawrence and Morata, 1977). For anterior and posterior wing the frequency of separation is very low, 0.7% (8/1200, Lawrence and Morata, 1977 and unpublished results). This suggests that anterior and posterior cells lie very close, if not adjacent, in the blastoderm. (3) At least during part of the larval period the anterior and posterior cells come into contact (see Bryant's fate map of mature wing disc, Bryant, 1975), and yet $M+$ clones never cross the antero-posterior boundary. By contrast when left and right discs of the same type come into contact during the embryonic or larval period, $M+$ clones can cross the left/right boundary; this has been shown for the first legs (Steiner, 1976), the genitalia (Dubendorfer, 1977), and the eye-antenna discs (Lawrence and Morata, unpublished). (4) As we shall discuss later, mutant cells for *engrailed* can transgress the antero-posterior compartment boundary, showing that the reason that normal cells respect the boundary is not physical separation.

B. *Larval Development*

The larval development of the wing, haltere and leg discs was studied by clonal analysis (Bryant and Schneiderman, 1969; Bryant, 1970; García-Bellido and Merriam, 1971a; Morata and García-Bellido, 1976). They all have a similar pattern of development that can be summarized as follows: Growth begins in the first larval period, starting with few cells (estimates ranging from 5-50 cells), and continues during all the larval period at a rate of 8-10 hours/cycle (at 25°C) until about 20h after pupariation when divisions stop as the cuticle is deposited. In the wing disc, early $M+$ clones (induced in embryos) are very large; but clones never cross the antero-posterior compartment boundary even though they may run along its entire length. This line runs in the wing blade between vein III and IV (Fig. 2), and in the thorax between

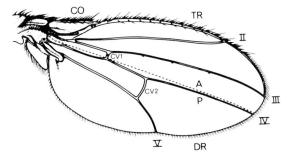

Fig. 2. Diagram of a normal living wing. The dotted line represents the boundary between anterior (A) and posterior (P) compartments. Notice the presence of sensilla on vein III and the first crossvein (CVS). CO costa, CV2 second crossvein, DR double row, TR triple row.

mesopleura and pteropleura, and notum and postnotum, and it is not signposted by any cuticular feature.

In the first instar anterior and posterior polyclones contain 10-20 and 5-10 cells, respectively (García-Bellido *et al.*, 1976; Lawrence and Morata, 1976). There are two more segregations later in the larval period which affect both anterior and posterior polyclones. They seem to take place at about the same time. The anterior and posterior poly-clones are each subdivided into two polyclones, one generating the dorsal part of the compartment and the other the ventral. Both groups of cells also become subdivided into daughter polyclones which generate the thoracic (proximal) and appendicular (distal) parts of anterior and posterior compartments. Thus after three segregations, the wing disc consists of eight polyclones, each differentiating a specific part of the adult cuticle (Fig. 3). As was pointed out by García-Bellido (1975), the

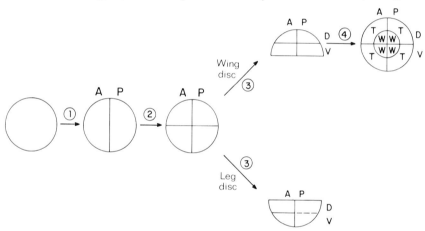

Fig. 3. Diagram of compartmentalization of the mesothoracic cuticular structures. Each subdivision is represented by a straight line. The first step (1) is the segregation of anterior (A) and posterior (P) polyclones, and this occurs probably at the blastoderm stage. The second (2) consists in the separation of the dorsal part of the segment (wing disc) and the ventral part (leg disc). This event takes place before 10 hours of development. During the larval period the wing disc becomes subdivided (3) into wing dorsal (D) and wing ventral (V), and about the same time (4) into thorax (T) and wing blade (W). For the leg disc there is evidence of the subdivision of the anterior compartment into dorsal and ventral (Steiner, 1976).

process of compartmentalization in the wing disc seems to progress in binary steps; every pre-existing polyclone is subdivided into two new ones by the same segregative event. In this way a few segregations can create many different compartments (n segregations making 2^n com-partments). In the wing disc, each compartment is a unique combination of the three elements of the binary formula; for example, ADT designates mesonotum (Fig. 3), but a change of dorsal to ventral produces

AVT which is mesopleura; a change of anterior by posterior produces PDT which is postnotum. There is some evidence that the three segregations described for the wing disc also take place in the haltere disc (Morata and García-Bellido, 1976) although for technical reasons (in certain regions of the haltere the marker mutants cannot be scored) the compartment boundaries cannot be defined precisely. The work of Steiner (1976) on the leg disc has shown that besides the subdivision into anterior and posterior compartments, at least the anterior compartment is further subdivided into dorsal and ventral parts (Fig. 3). The genitalia also become subdivided during later development (Dubendorfer, 1977).

IV. GENES INVOLVED IN COMPARTMENTALIZATION

We have seen that there are at least four different segregative steps during the development of the thoracic cuticle. *A priori* it is difficult to predict what the phenotype of mutations blocking or interfering with any of these steps might be, especially if each of them is in itself a complex genetic process involving more than one gene. Probably mutations in some of these genes will result in homologous effects in several discs because some of the segregations are common to more than one disc. We know of some mutations in which one compartment is transformed into another of the same disc, causing duplication of one at the expense of another. In *engrailed* the posterior compartments of wings, legs, and halteres (García-Bellido and Santamaria, 1972; García-Bellido *et al.*, 1973, 1976) are partially transformed into the corresponding anterior ones. In *wingless* the distal compartments (appendages) of wings and haltere discs are replaced by their respective thoracic counterparts (Sharma and Chopra, 1976). These phenotypes suggest an effect of *engrailed* on the antero-posterior segregation and of *wingless* on the thorax/appendage segregation.

A. *Engrailed and the Segregation of Anterior and Posterior Polyclones*

The antero-posterior boundary divides the wing blade into anterior and posterior compartments which have a different overall pattern although most of the local differentiation is identical. The pattern of veins and sensilla and bristles (Fig. 2) is specific for each compartment. Differences in anterior and posterior compartments of the legs have also been described (Steiner, 1976). The mutation *engrailed* changes the posterior wing compartment into a partial mirror image of the anterior one (García-Bellido and Santamaria, 1972). The line of mirror image symmetry runs at or close to the normal position of the antero-posterior compartment boundary (Fig. 4). A similar transformation is found in the legs where, for instance, the sexcomb of the anterior male foreleg is duplicated in the posterior male foreleg (Tokunaga, 1961).

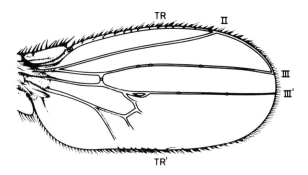

Fig. 4. Diagram of an *engrailed* wing. The posterior margin shows the presence of bristles (TR') typical of the anterior margin (TR). Note that a vein (III'), in a similar position to vein IV in wildtype wings, bears sensilla.

These observations immediately posed the question of whether or not there exists a normal antero-posterior compartment boundary in *engrailed* wings. This question was answered by producing marked clones using both conventional clonal analysis and the *Minute* technique; these clones showed that it is not possible to define a normal antero-posterior compartment boundary in *engrailed* wings (Morata and Lawrence, 1975; Lawrence and Morata, 1976). Most of the clones tended to be restricted to either the anterior or to the posterior part of the wing, indicating that there are originally two separate groups of cells which can intermingle near the middle of the wing. Thus the complete segregation of anterior and posterior polyclones is dependent on the normal function of *engrailed*.

The question of the maintenance of the antero-posterior boundary and its dependence on the local function of the *en+* gene in the cells of the borderline was studied in a series of experiments in which *Minute+ en/en* clones (marked with *pawn*, a bristle and trichome marker) were produced in *Minute en/+* (phenotypically wildtype) wings. It was found that *en/en* clones fell into two classes: (i) those which were exclusively anterior had no effect on the normal anterior pattern, and if they reached the central region of the wing, defined the normal antero-posterior boundary, and (ii) clones that were mainly posterior had the typical *engrailed* phenotype, and if they reached to the central region of the wing, frequently crossed to territory occupied by anterior cells (Lawrence and Morata, 1976). The conclusion from these experiments is that the function of *en+* is needed by the posterior cells to maintain the compartment boundary, while *en+* function is immaterial for the anterior cells. This dependence lasts until late development; even late posterior clones can cross into anterior territory.

The effect of *engrailed* on the posterior wing pattern was studied by producing *en/en* clones in wildtype wings. It was observed that *en/en* clones showed *engrailed* phenotype autonomously; that is, they differentiated anterior structures. This behaviour was found even for late small clones (García-Bellido and Santamaria, 1972; Lawrence and Morata, 1976) indicating that the normal posterior pattern is dependent on *en+* activity until late development. By contrast *en/en* clones showed no effect in the anterior region of the wing.

In summary, the wildtype function of *engrailed* seems to be restricted to the posterior compartment where it is responsible for the developmental characteristics of the posterior cells. One of its functions is to prevent posterior cells from mixing with anterior cells along the compartment border. It has been suggested (Morata and Lawrence, 1975) that *en+* "labels" the surface of posterior cells which thereby acquire specific cell affinities. Indeed, after dissociation and mixing, wildtype anterior and posterior cells will sort out from each other, but *engrailed* posterior cells will intermingle with both anterior and posterior wildtype cells (García-Bellido and Santamaria, 1972).

We should point out that not all aspects of *engrailed* are wholly consistent with an effect on the segregation of anterior and posterior polyclones. The conflicting evidence comes from an analysis of the extent of the transformation. In the wing disc, where it has been studied in detail, only the posterior wing compartment is transformed (partially) into anterior, whereas the posterior thoracic compartments are not. One should expect the postnotum (posterior dorsal) and ptero-pleura (posterior ventral) to be transformed into mesonotum (anterior dorsal) and mesopleura (anterior ventral) in mirror image fashion. The fact is that these compartments are not affected. There are several possible explanations; the one we favour is that *engrailed* is leaky, the expressivity of the mutation being very low in the thoracic compartments. This may be a necessary condition for viability since transformations affecting all thoracic parts are very likely to be lethal. It is known that some homeotic mutants have very different expression in different compartments (Morata, 1975).

B. Wingless and the Segregation of Thoracic Appendages in Wings and Halteres

The mutation *wingless* was discovered recently (Sharma and Chopra, 1976). It is a recessive mutation that has a complex phenotype. In the majority of *wingless* discs the appendicular parts of wing and haltere discs are replaced by a mirror image duplication of the thoracic derivatives of the same discs. In the remainder of the discs, the phenotype is normal. This all-or-none phenotype is different from other homeotic mutations where, typically, there is a continuous range of transformations from more to less extreme. Another difference from other homeotic mutants so far tested is that marked clones of cells homozygous for *wingless* do not show the transformation (Morata and Lawrence, 1977b). These facts suggest that *wingless* affects the determination of groups of cells so that the mirror image duplications of meso and meta-notum could result from some interference with the normal segregation of thoracic and appendicular primordia. There is, however, another explanation: Bryant (1975) found that thoracic (and other) fragments of the wing disc can duplicate when they are isolated and cultured *in vivo*. Based on this observation, the phenotype of *wingless* can be explained as the result of localized *cell death* in the distal region of the wing disc (which is presumptively the wing blade and pleuras), followed by growth and duplication of the remaining thoracic fragment (Sharma and Chopra, 1976). Presumably the same process would take place in the haltere disc.

In a series of experiments we tried to discriminate between these two explanations for the phenotype of *wingless* (Morata and Lawrence, 1977b). We found some cases of *wingless* gynandromorphs in which the duplicated notum was a mosaic of male and female cells, while the normal notum was entirely male. This seems incompatible with the cell death hypothesis since there is no way in which male cells can produce female cells in gynandromorphs. Other evidence against the cell death hypothesis comes from the clonal analysis of the development of the *wingless* discs during the larval period (Morata and Lawrence, 1977b). The frequency and size of clones in normal and duplicated nota were the same throughout the entire larval period; most of the clones were restricted to only one of the two nota. These results indicate that the normal and duplicated nota have independent lineages and derive from approximately the same number of founder cells.

However, although we believe that these results rule out the cell death hypothesis, the relationship between *wingless* and the process of compartmentalization is not clear. There are some features of the *wingless* phenotype that apparently do not fit with an effect on compartments. One example is the location of the mirror plane in the duplication, which in most cases is close to where the thorax joins the wing. But sometimes the mirror plane can be so distal as to include part of the tegula (which is thought to be part of the wing compartment (García-Bellido *et al.*, 1976)) or very proximal, so that the two nota are symmetrically reduced to vestiges. Some of this variation in phenotype could be explained by our observation that the mutant produces a band of cell death along the mirror plane; however this must be a secondary consequence of the duplication because, as we have explained, massive cell death cannot be the cause of the phenotype. Moreover, unlike the antero-posterior and dorso-ventral compartment border, the notum-wing "compartment border" is very difficult to locate precisely (Lawrence and Morata, unpublished observations).

In summary, our results show that the two nota begin life as two groups of cells, one of which probably derives from those cells which in wildtype wing discs (or *wingless* discs showing wildtype phenotype) give rise to wing. In our view, *wingless* results in a double group of cells (the anterior and posterior presumptive wing) taking an abnormal developmental route: the non-autonomy of *wingless* clones suggests that this effect occurs early, and afterwards is not influenced by the function of the *wingless* gene.

V. THE MODE OF ACTION OF HOMEOTIC GENES

Several homeotic mutants produce transformations that are defined by compartment boundaries. Apart from the cases of *engrailed* and *wingless,* which result in intradisc transformations, other mutants produce intersegmental transformations. For example, in an extreme *bithorax* phenotype, the anterior metathoracic compartments are transformed into anterior mesothoracic compartments. Thus the anterior haltere is transformed into anterior wing and anterior third leg is transformed into anterior second leg, the limit of the change coinciding exactly with the antero-posterior compartment boundary as defined by the *Minute* technique. Extreme *post-bithorax* phenotypes show the posterior metathoracic compartments (leg and haltere) transformed into posterior mesothoracic ones. The limit of this transformation is also at the antero-posterior boundary. These results suggest that the activities of the wildtype genes for *bithorax* and *post-bithorax* are restricted to the anterior and posterior polyclones respectively. As we have seen, the wildtype gene for *engrailed* is also restricted to the posterior polyclones of (at least) the legs, wings, and halteres.

These observations have led to the general hypothesis that polyclones are the developmental units which are controlled by the homeotic genes. In most homeotic genes the activity of the wildtype allele is continually required to maintain the developmental route of the polyclone: García-Bellido (1975) has called such genes "selector genes". The function of selector genes interact additively: for example, the *engrailed* gene determines posterior development in wings, legs and halteres, but the posterior haltere and third leg also require the function of the wildtype allele of the *postbithorax* gene; in the case of the double mutant combination *engrailed; postbithorax*, the posterior haltere develops as anterior wing (García-Bellido and Santamaria, 1972). In this view each compartment is specified by a unique combination of selector genes. One could speculate that there are other selector genes, yet to be discovered in mutant form, that control the segregation of, for instance, the dorsal from the ventral part of the segment. Such a mutant would transform, for example, the second leg into wing (or viceversa), and the third leg into haltere. A mutant of this kind would probably be lethal, which may explain why it has not been found.

We know very little of the processes that define polyclones and activate the selector genes. The precise restriction of selector genes to particular polyclones suggests that the polyclones are defined *before* the genes become activated. There are thus at least two steps: (i) the definition of groups of cells, (ii) the activation of the appropriate selector genes in each group. The first step appears to be geographic (Crick and Lawrence, 1975), cells being allocated according to their position not their ancestry.

Certain experimental treatments applied at blastoderm produce *bithorax* phenocopies (Gloor, 1947), and it has been shown (Capdevila and García-Bellido, 1974) that these phenocopies are the result of a permanent inactivation of the *bithorax* gene in some cells. This suggests that the phenocopy treatment interferes with the second step — activation of the *bithorax* gene. After this step, the activity of selector genes is thought to be continuous until the completion of development (Morata and García-Bellido, 1976; Morata and Lawrence, 1977a).

In certain cases a given polyclonal segregation can be followed by the activation of more than one selector gene. In the metathoracic segment, the gene $bx+$ is activated in the anterior polyclone and the genes $en+$ and $pbx+$ are activated in the posterior one, all those events possibly occurring simultaneously.

However, it is not yet clear if all homeotic mutants produce transformations on a compartment basis. In this respect, the transformations produced by *spineless-aristapedia* or *Antennapedia* (see Lindsley and Grell, 1968) are particularly intriguing; part of the eye-antenna is apparently transformed into a complete mesothoracic leg. This homeotic leg presumably possesses anterior and posterior compartments as normal legs do (Steiner, 1976). However, the *Minute* technique does not reveal any early compartment boundary in the wildtype eye-antenna structures; even late in development *Minute+* clones can cross from eye to antenna (Lawrence and Morata, unpublished). Yet antenna cells alone can develop, if they are mutant for an extreme antenna-leg transformation like *Nasobemia* (Gehring, 1966), a complete mesothoracic leg. Furthermore *aristapedia* eye cells, that do not suffer the transformation "in situ", can regenerate a homeotic leg (Gehring and Schubiger, 1975). A similar situation occurs in *bithoraxoid* where the first abdominal segment is transformed into a thoracic one consisting of anterior metathorax and posterior mesothorax. The reason for this singular phenotype is that mutants in the *bithoraxoid* locus result in two transformations (Lewis, 1963, 1964), one attributable specifically to the *bxd* locus which transforms the first abdominal segment into metathorax, and the other a *postbithorax* transformation that changes posterior metathorax into posterior mesothorax. In extreme cases it produces an extra (fourth) pair of legs in place of the sternite, and a haltere-wing appendage and thoracic tissue in place of the tergite. These homeotic structures contain anterior and posterior compartments although, probably, abdominal histoblasts are not subdivided into anterior and posterior compartments (Roseland and Schneiderman, unpublished). However, a recent observation (Lawrence *et al.*, 1977) that the frequency of mosaicism for the first tergite is double that of the other tergites may indicate that this particular segment might be a special case and contain both anterior and posterior polyclones.

REFERENCES

Anderson, D. T. (1963). *J. Embryol. Exp. Morphol.* **11**, 339-351.

Becker, H. J. (1957). *Z. indukt. Abstamm. U. Vererb. L.* **88**, 333-373.

Bryant, P. J. (1970). *Develop. Biol.* **22**, 389-411.

Bryant, P. J. (1975). *J. Exp. Zool.* **193**, 49-78.

Bryant, P. J. and Schneiderman, H. A. (1969). *Develop. Biol.* **20**, 263-290.

Capdevila, M. P. and Garcia-Bellido, A. (1974). *Nature New Biol.* **250**, 500-502.

Crick, F. H. C. and Lawrence, P. A. (1975). *Science* **189**, 340-347.

Dubendorfer, K. (1977). Thesis, University of Zurich.

Ferris, G. F. (1950). *In:* "Biology of *Drosophila*" (M. Demerec ed.) pp. 368-418, New York.

Ferrus, A. (1975). *Genetics* **79**, 589-599.

García-Bellido, A. (1975). *In:* "Cell Patterning". Ciba Foundation Symposium **29**, 161-182. North Holland.

García-Bellido, A. and Ferrus, A. (1975). *Wilhelm Roux' Arch.* **178**, 337-340.

García-Bellido, A. and Merriam, J. R. (1969). *J. Exp. Zool.* **170**, 61-75.

García-Bellido, A. and Merriam, J. R. (1971a). *Develop. Biol.* **24**, 61-87.

García-Bellido, A. and Merriam, J. R. (1971b). *Develop. Biol.* **26**, 264-276.

García-Bellido, A., Ripoll, P. and Morata, G. (1973). *Nature New Biol.* **245**, 251-253.

García-Bellido, A., Ripoll, P. and Morata, G. (1976). *Develop. Biol.* **48**, 132-147.

García-Bellido, A. and Santamaria, P. (1972). *Genetics.* **72**, 87-104.

Gehring, W. J. (1966). *Arch. Julius Klaus-Stift.* **41**, 44-54.

Gehring, W. J. (1976). *Ann. Rev. Genet.* **10**, 209-252.

Gehring, W. J. and Nothiger, R. (1973). *In:* "Developmental Systems: Insects" (S. Counce and C. H. Waddington eds.) pp 211-290. Academic Press, New York.

Gehring, W. J. and Schubiger, G. (1975). *J. Embryol. Exp. Morphol.* **33**, 459-469.

Gloor, H. (1947). *Rev. Suisse Zool.* **54**, 637-712.

Hotta, Y. and Benzer, S. (1972). *Nature.* **240**, 527-535.

Janning, W. (1974a). *Wilhelm Roux' Arch.* **174**, 313-332.

Janning, W. (1974b). *Wilhelm Roux' Arch.* **174**, 349-359.

Janning, W. (1976). *Wilhelm Roux' Arch.* **179**, 349-372.

Kankel, D. R. and Hall, J. C. (1976). *Develop. Biol.* **48**, 1-24.

Lawrence, P. A., Green, S. and Johnston, P. (1977). *J. Embryol. Exp. Morphol.* (in press).

Lawrence, P. A. and Morata, G. (1976). *Develop. Biol.* **50**, 321-337.

Lawrence, P. A. and Morata, G. (1977). *Develop. Biol.* **56**, 40-51.

Lewis, E. B. (1963). *Am. Zool.* **3**, 33-56.

Lewis, E. B. (1964). *In:* "The Role of Chromosomes in Development" (M. Locke ed.) pp 231-252. Academic Press, New York.

Lindsley, D. L. and Grell, E. H. (1968). *Carnegie Inst. Wash. Publ.* No. 627.

Morata, G. (1975). *J. Embryol. Exp. Morphol.* **34**, 19-31.

Morata, G. and Garcia-Bellido, A. (1976). *Wilhelm Roux' Arch.* **179**, 125-143.

Morata, G. and Lawrence, P. A. (1975). *Nature.* **225**, 614-617.

Morata, G. and Lawrence, P. A. (1977a). *Nature.* **265**, 211-216.

Morata, G. and Lawrence, P. A. (1977b). *Develop. Biol.* **56**, 227-240.

Morata, G. and Ripoll, P. (1975). *Develop. Biol.* **42**, 211-221.

Postlethwait, J. H. and Schneiderman, H. A. (1971). *Develop. Biol.* **25**, 606-640.

Ripoll, P. (1972). *Wilhelm Roux' Arch.* **169**, 200-215.

Ripoll, P. and Garcia-Bellido, A. (1973). *Nature New Biol.* **241**, 15-16.

Sharma, R. P. and Chopra, V. L. (1976). *Develop. Biol.* **48**, 461-465.

Steiner, E. (1976). *Wilhelm Roux' Arch.* **180**, 9-30.

Stern, C. (1936). *Genetics.* **21**, 625-630.

Struhl, G. (1977). (in preparation).

Sturtevant, A. H. (1929). *Z. wiss. Zool.* **135**, 323-356.

Tokunaga, C. (1961). *Genetics.* **46**, 158-176.

Wieschaus, E. and Gehring, W. (1976a). *Wilhelm Roux' Arch.* **180**, 31-46.

Wieschaus, E. and Gehring, W. (1976b). *Develop. Biol.* **50**, 249-263.

II. Vertebrates

Pattern Regulation and Cell Commitment in Amphibian Limbs

Susan V. Bryant

Center for Pathobiology
and
Department of Developmental and Cell Biology
University of California
Irvine, California 92717

I. INTRODUCTION

The secondary embryonic fields of many animals, both vertebrate and invertebrate, are capable of extensive pattern regulation at certain stages during the lifetime of the animal. This response is evoked when the normal relationships between the constituent cells are disturbed, such as when parts of the field are removed, added or reoriented. Pattern regulation in secondary fields does not always lead to the restoration of a normal pattern, and the response to a disturbance may rather be to duplicate the parts already present. However, it has recently been shown that the different types of response which occur in secondary fields are similar in very different animals, and that they can be understood in terms of simple rules for cellular behavior, which are the same whether the field in which pattern regulation is occurring is the leg of a cockroach, an imaginal disc of *Drosophila* or the limb of a newt (French *et al.*, 1976; P. Bryant, *et al.*, 1977). In each of these organs, pattern regulation is dependent upon cell division, and new pattern elements are added on

63

during growth. That is, in the terminology introduced by Morgan (1901), these three fields behave epimorphically. French *et al.* (1976), in considering these three epimorphic fields, which are basically either cone or disc shaped, have pointed out that the cells behave as if they had information about their position on a circle or circumference, and their position on a radius. Pattern regulation occurs when normally non-adjacent cells are brought together by grafting or during wound healing. It is assumed that growth is stimulated at discontinuities in either the circular or radial sequences of positional values, and that cells with positional values which are intermediate between those of the opposed values at the discontinuity are intercalated. In addition, it is proposed that the characteristic "distal transformation" which occurs in epimorphic fields (Rose, 1962) can take place only when a complete set of positional values in the circular sequence is present. The proposed arrangement of positional values in an epimorphic field is shown in Fig. 1. In the amphibian limb, the distal tip of the appendage would be at the field center, and the proximal boundary of the limb field would correspond to the outer circle. The proximal-distal axis of the limb is therefore represented by the radial positional values, A-E.

The evidence from cockroaches, *Drosophila* and amphibians which led to the development of this polar coordinate model are presented in the paper by French *et al.* (1976). In this article, the various regulative responses of the regenerating limbs of amphibians to different types of surgical intervention will be discussed in terms of this interactive model.

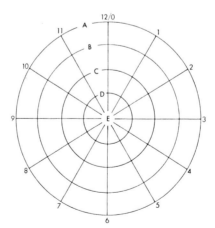

Fig. 1. Polar coordinates of positional information in an epimorphic field. Positional values are arranged in circular and radial sequences and for convenience, only 12 values in the circular sequence (0-12) and five in the radial sequence (A-E) are shown. In the amphibian limb, the proximal boundary of the field is at level A and the distal tip at level E. (From Bryant and Iten, 1976.)

In addition, the relationship between the formal model and the properties of different tissues of the limb will be explored. In particular, evidence concerning the location of positional information will be considered, as will the degree to which different cells of the limb are bounded by previously acquired heritable commitments.

II. DISTAL TRANSFORMATION

Pattern regulation leading to the formation of the distal parts of an appendage has been termed distal transformation (Rose, 1962). When an amphibian limb is transected at any level along the proximal-distal axis, the cut surfaces of both the proximal stump, and the distal, amputated portion are capable of distal transformation to form all of the structures which normally lie distal to the level of the cut. In the case of the proximal

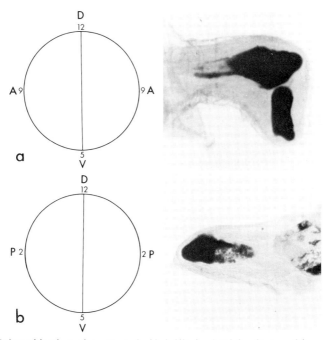

Fig. 2. Failure of distal transformation in double-half limbs. a) At left, a diagram of the composition of a double-anterior limb stump with key positional values indicated. At right, a whole mount preparation of a double-anterior limb, 2-1/2 months after amputation. A separate, darkly staining skeletal element has formed at the distal tip, and new cartilage has been added to the end of the humerus. X 14. b) At left, a diagram of the composition of a double-posterior limb stump. At right, a whole mount preparation of a double-posterior limb two months after amputation. No new skeletal elements have formed, but cartilage has been added to the distal tip of the humerus. X 16. D=dorsal; V=ventral; A=anterior; P=posterior. (Parts from Bryant, 1976).

stump, distal transformation leads to regeneration of the missing parts. The amputated portion must be grafted so as to ensure an adequate vascular and nervous supply to the cut surface. This can be achieved by grafting the distal tip of the limb either into the flank (Dent, 1954; Butler, 1955; Deck, 1955) or onto the contralateral limb stump (Carlson, *et al.*, 1974) prior to amputation. Under these conditions, distal transformation from the originally proximal-facing cut surface results in a duplication of the distal structures.

French *et al.* (1976) have suggested that complete distal transformation, not only in amphibian limbs, but in other epimorphic fields as well, can take place only from a surface at which a complete set of positional values in the circular sequence is present. In order to test the validity of this proposal for the newt *Notophthalmus viridescens,* upper arm stumps which were normal in basic tissue composition, but which consisted of a double set of only half of the circumferential positional values were made by grafting (Bryant, 1976; Bryant, 1977; Bryant and Baca, 1978).

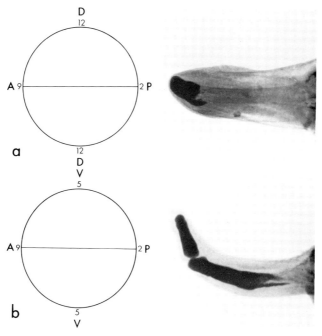

Fig. 3. Failure of distal transformation in double-half limbs. a) At left, a diagram of the composition of a double-dorsal limb stump. At right, a whole mount preparation of a double-dorsal limb, two months after amputation. Cartilage has been added to the tip of the humerus. X 14. b) At left, a diagram of the composition of a double-ventral limb stump. At right, a whole mount preparation of a double-ventral limb, two months after amputation. A single new skeletal element has formed. X 13. Abbreviations as in Fig. 2.

The majority of such double-half limbs, whether double-anterior, double-posterior, double-dorsal or double-ventral, failed to undergo normal distal transformation (Figs. 2 and 3) although some new tissue was regenerated. Similar experiments with similar results, have now been completed in axolotls (Tank, 1978b). Control limb stumps, in which a longitudinal half of the upper arm had been removed and replaced, regenerated normally. These results strongly indicate the need for a complete circumference of positional values in order for complete distal transformation to occur. This conclusion is strengthened by the finding that some of the double-half limbs formed supernumerary limbs at the proximal edge of the grafted tissue. At this location, a complete circle of positional values is created by the juxtaposition of host and graft tissues (Fig. 4). In the control, sham-operated limbs such a complete circle is not generated at the base of the graft, and in no case did a supernumerary limb develop in this location.

Numerous studies of amphibian limb regeneration have shown that in order for distal transformation to occur, certain local conditions at the amputation site must be met. These conditions are an adequate nerve supply, a thickened wound epidermis, and an accumulation of dedifferentiated cells to form a blastema (Thorton, 1968; Carlson, 1974a; Singer, 1974; Tassava and Mescher, 1975). In the past, it has been generally assumed that these basic conditions are not only necessary, but also sufficient for distal transformation from a limb stump of normal tissue composition. However, an analysis of the double-half limbs now shows this assumption to be incorrect. The amputation site of double-half

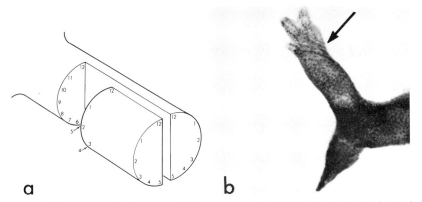

Fig. 4. a) Diagram of a double-posterior limb stump on a right limb. Positional values in the circular sequence are indicated on the graft and host. The slightly unequal spacing of positional values was suggested by previous experiments (Bryant and Iten, 1976). At the proximal edge of the graft, a complete circle of positional values is created at the graft/host junction. b) Double-posterior limb stump in which a supernumerary limb (arrow) formed at the proximal graft/host junction. Amputation through the distal region of the graft led to the formation of an abortive outgrowth. X 10. (From Bryant, 1977.)

limbs is well innervated (Fig. 5) and a comparison between double-half limbs and their appropriate sham-operated controls shows no correlation between the quanitity of innervation and the completeness of distal transformation (Fig. 6). In addition, all double-half limbs initiate regeneration, and during the early stages are indistinguishable from controls. As can be seen from Fig. 7, a thickened wound epidermis and a sizeable accumulation of blastema cells are present on double-half limb stumps. Furthermore these blastemas, which often only develop to

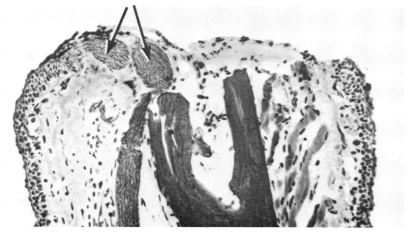

Fig. 5. A longitudinal section of a double-posterior limb stump at the time of amputation. The section was stained with Bodian's silver stain for nerve fibers. The arrows indicate the bundles of viable nerve fibers at the amputation plane. X 90.

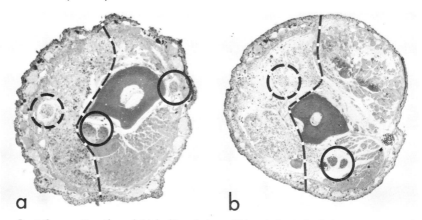

Fig. 6. Cross sections through (a) double-anterior and (b) control-anterior limb stumps, one month after grafting. Grafted tissue is to the left of the dotted lines, and in both sections some disorganization of the muscles is apparent. The bundles of nerve fibers which remained undisturbed during grafting are ringed by solid circles; those which were cut are ringed with broken circles. X 36. (From Bryant, 1977.)

medium or late bud stages, appear to be normally innervated (Fig. 7). Hence, the failure of double-half limbs to undergo normal distal transformation cannot be attributed to a failure to develop either an adequate innervation, a thickened wound epidermis or a blastema. Therefore, it must be concluded that these conditions, though necessary are not sufficient for distal transformation.

One of the obvious questions which emerges from the suggestion of the requirement for a complete circumference, is just how the circumference is involved in the generation of all the more distal limb levels. Although it has been clearly established that during distal transformation in vertebrate limbs new pattern elements are differentiated in a proximal to distal sequence, the order in which these different elements are actually specified is not yet known. Stocum (1975a), Faber (1976) and Maden (1977) have proposed that the proximal and distal positional values of the regenerate (representating low and high points of a gradient in positional information) are established first, with a initially steep gradient between them. The gradient flattens out by

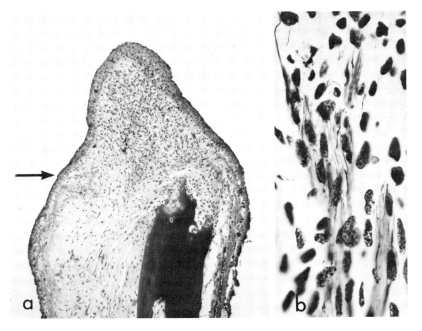

Fig. 7. a) A longitudinal section of a double-posterior limb two weeks after amputation. An apparently normal regenerate at the stage of medium bud has formed. The arrow is on the grafted side of the limb and indicates the approximate level of amputation. Mallory's triple stain. X 40. (From Bryant, 1977.) b) Nerve fibers among the cells of a double-posterior blastema. Bodian's silver stain. X 350.

growth and the pattern elements differentiate in a proximal-distal sequence. Hence, according to this theory, positional values are not established in a proximal-distal sequence, although they are expressed in this sequence. A sequential model for the specification of pattern elements during distal transformation in developing chick limbs has been proposed by Summerbell *et al.* (1973) and it has also been applied to regenerating amphibian limbs by Smith *et al.* (1974) and Smith and Crawley (1977). According to this model, the specification of pattern elements occurs in a proximal-distal sequence which is reflected by the sequence of differentiation. At the tip of the limb bud or blastema, cells are in a "progress zone". The amount of time spend (or possibly the number of cell cycles traversed) by cells in the progress zone determines their proximal-distal positional value. Thus, cells which leave the progress zone early in distal transformation will have more proximal positional values than ones which leave later.

Although no experiments performed to date can clearly distinguish between these two different ideas it is interesting to consider them in relation to the requirement for a complete circumference of positional values. If in amphibian regeneration, a complete circumference is required to establish the distal boundary in the models of Stocum (1975a), Faber (1976), and Maden (1977), then double-half limbs should not be able to generate the distal boundary, and would not be expected to produce any new structures beyond the level of amputation. However, as we have seen, double-half limbs do initiate regeneration, and do form some new structures. Alternatively, if the circumference at any one level is involved in the generation of the next most distal level, then the result would be a proximal-distal sequence of establishment of the pattern in normal limbs, and some distal transformation in double-half limbs. One way in which a circumference at one level might generate the next most distal circumferential level is illustrated in Fig. 8. It is proposed that dedifferentiated cells move in toward the center of the wound epithelium, and in so doing, come into contact with cells from which they were previously separated by a short arc around the circumference. Intercalation by the shortest route between these cells will generate a second complete circumference, level C in Fig. 8. It is further proposed that these newly generated cells will occupy the most distal location beneath the wound epithelium. As the epidermis grows, new space could be made available at the very tip of the mesoderm as suggested by Summerbell and Wolpert (1972). Interaction between the cells in the newly generated level C will lead to the formation of a level D circumference, and so on until the most distal level has been formed. In double-half limbs, a limited amount of intercalation, as shown in Fig. 9 could occur before all positional discontinuities were resolved, but

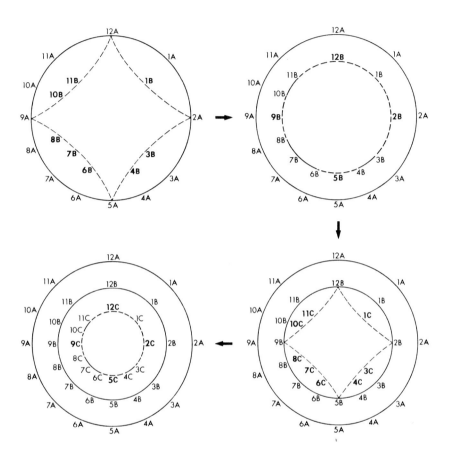

Fig. 8. Diagram to illustrate how a complete circumference of positional values at limb level A could generate successively more distal limb levels by circumferential intercalation. The slightly unequal spacing of positional values was suggested by previous experiments (Bryant and Iten, 1976). a) Immediately after amputation, it is proposed that mesodermal cells dedifferentiate and migrate towards the center of the newly healed wound epithelium. Cells which were previously separated from each other will now be brought into contact. In this diagram only the cells with positional values 12, 2, 5, and 9 are shown as interacting for simplicity. Shortest intercalation between these interacting cells generates cells with the new positional values of level B, shown in bolder type. b) The newly formed cells take up the most distal location under the wound epithelium, and now cells with the remaining positional values (indicated in bolder type) in the level B circumference are intercalated. c) As space becomes available at the distal tip of the limb due to growth of the epithelium, cells in the level B circumference which are in contact with each other across a short arc of the circumference, intercalate to form cells of level C. As in a), only cells with values 12, 2, 5, and 9 are shown as interacting for simplicity. The newly formed cells are indicated in bolder type. d) As in b), cells with the remaining positional values in the level C circumference are filled in by intercalation, and are indicated in bolder type. This process will continue until level E, the field center, is reached.

intercalation would cease before the most distal structures were formed. This would result in the appearance of normal, early stages of regeneration in these limbs; further development would cease when no more distal values could be generated by intercalation. Hence, the results

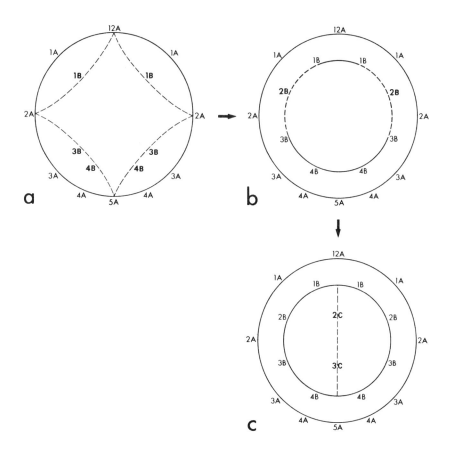

Fig. 9. Diagram to illustrate the formation of distally incomplete regenerates from double-half limbs by circumferential intercalation. It is proposed that the dedifferentiated mesodermal cells behave in the same way as that described in Fig. 8 for normal distal transformation. Again, as in Fig. 8, the slightly uneven spacing of positional values was suggested by previous experiments (Bryant and Iten, 1976). a) Interaction between previously separated cells of limb level A will lead to the intercalation of cells with level B positional values (shown in bolder type). As in Fig. 8, only cells with positional values 12, 2, 5, and 9 are shown as interacting for simplicity. b) Intercalation between the cells of level B will generate some new values (shown in bolder type) in level B, but there will not be as many circumferential positional values in level B as in level A. c) Intercalation between the cells of level B will generate new values (shown in bolder type) which satisfy all positional disparities, and distal transformation will come to a halt.

obtained with double-half limbs are consistent with a model of distal transformation in which the new pattern elements are established in a proximal-distal sequence. The requirements for a complete circumference of positional values can be understood in terms of its proposed role in the establishment of new more distal circumferences by shortest intercalation, and the region of the limb in which this intercalation is occurring can be considered to be equivalent to the progress zone of Summerbell *et al.* (1973).

III. INTERCALARY REGENERATION

Experiments to demonstrate intercalation in both the proximal-distal and circumferential sequences of positional values have been carried out in amphibians. When a regenerate from a distal limb level is transplanted to a more proximal stump of either the same or the contralateral limb, intercalary regeneration between graft and stump occurs and the resulting limb is complete in the proximal-distal axis (Fig. 10). This type of intercalation in the proximal-distal sequence has been shown in newts (Iten and Bryant, 1975; Bryant and Iten, 1977) and in salamanders (Stocum. 1975b; Tank, 1978a). Both the graft and stump tissues undergo extensive dedifferentiation prior to intercalation (Iten and Bryant, 1975; Stocum, 1975b). Additional experiments by Stocum (1977) indicate that complete intercalary regeneration does not occur without nerves and epidermis. The ability to replace all missing pattern elements when the defect includes part of two limb segments, has also been shown to decrease in advanced regenerates of salamanders (Stocum, 1975b) and in the mature limbs of newts (Bryant and Iten, 1977). In the latter case, intercalation of the missing distal portion of the upper arm segment occurs, but the lower arm segment is not replaced.

Intercalation in the circumferential sequence of positional values has been inferred from the results of blastema grafts which lead to the formation of supernumerary limbs (Iten and Bryant, 1975; Bryant and Iten, 1976; Tank, 1978a). If a regenerate is transplanted to a contralateral limb stump with either anterior and posterior or dorsal and ventral circumferential positions of graft and stump opposed, supernumerary limbs arise from the graft junction. These limbs have consistent locations, orientations and handedness, all of which can be predicted by shortest intercalation between opposed positional values at the graft junction (Fig. 11). The supernumerary circles form the bases from which distal transformation proceeds to form all of the more distal levels of the supernumerary limbs. As with distal transformation following simple amputation, it is presumed that supernumerary distal transformation will be possible only if the usual local requirements for nerves, epidermis and blastema formation are satisfactory.

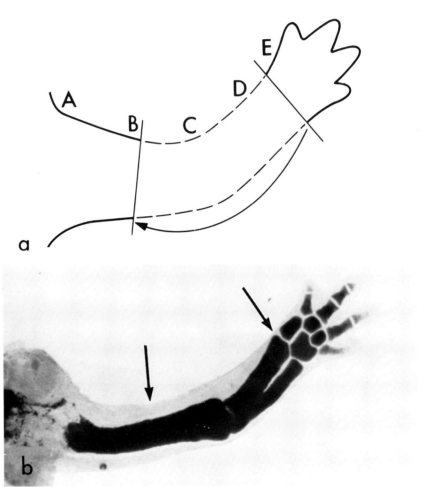

Fig. 10. Intercalation in the proximal-distal sequence in the newt limb. a) Diagram to illustrate the grafting procedure. An advanced regenerate originating at a distal limb level (level E) was grafted ipsilaterally and without rotation to an upper arm stump (level B). b) Skeletal preparation of a limb treated as described in a) three months after grafting. Those limb levels deleted by grafting have been intercalated, giving rise to a normal limb. Arrows indicate estimated locations of original graft and stump. X 16. (From Bryant and Iten, 1977.)

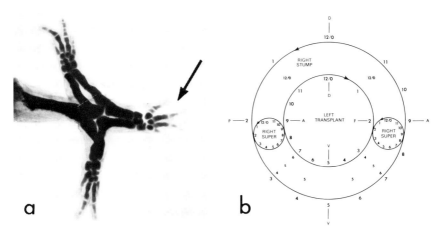

Fig. 11. The formation of supernumerary limbs in newts. a) Skeletal preparation of a right limb onto which was grafted an advanced regenerate from the left limb in place of its own regenerate. Anterior and posterior positions of graft and stump were apposed in grafting. The grafted regenerate developed a left hand (arrow), and in the graft junction, anterior and posterior supernumerary limbs of right handedness have formed. X 9. b) Diagrammatic interpretation of supernumerary limb formation as shown in a) according to the polar coordinate model. Outer circle is host circumference; inner circle graft circumference. The slightly unequal spacing of positional values was suggested by other experiments (Bryant and Iten, 1976). The numbers in the graft junction are positional values intercalated by the shortest route between opposed host and graft positional values. The shorter route around the circle is different on each side of the points of maximum incongruity, so a complete circle of positional values is generated at those positions. Distal transformation occurs from these complete circles to generate supernumerary limbs, whose orientation and handedness is determined by the direction of intercalation in the adjacent regions of the graft junction. The direction of intercalation is a consequence of shortest intercalation. As can be seen, the supernumerary limbs have right handedness and are oriented in the same fashion as the stump. Abbreviations as in Fig. 2. (From Bryant and Iten, 1976.)

IV. LOCATION OF POSITIONAL INFORMATION

In the preceding sections, we have seen that the polar coordinate model of French *et al.* (1976) is capable of explaining a variety of complex morphogenetic events in amphibian limb regeneration in terms of two simple rules for cellular behavior. Since this model assumes that positional information is specified in a two-dimensional array of cells, yet the amphibian limb is a three-dimensional structure we must now consider a further suggestion; that positional information is a property of only one tissue layer, and that the other tissue layers become specified for pattern formation secondarily by contact with the first layer. The tissue layers of the limb are epidermis, connective tissue of the dermis, connective tissue of the muscles, muscle, and bone or cartilage. From the evidence presented below, it seem likely that positional values are established in the connective tissue component of the limb.

It has been known for many years that X-irradiated limbs are not regenerated following amputation (Brunst, 1950). It has recently been confirmed that this inhibition is due to a failure of the cells in the limb to successfully complete mitosis (Maden and Wallace, 1976). However, if the skin of an irradiated limb is removed and replaced with a cuff of skin from around the circumference of an unirradiated limb, normal regeneration can occur in a high percentage of the cases (Carlson, 1974b; Lheureux, 1975b). Furthermore an unirradiated cuff of skin which was derived from only one portion of the circumference (e.g., a longitudinal strip of dorsal skin turned through 90° and wrapped around the limb), was in most cases unable to stimulate regeneration of X-irradiated limbs (Carlson, 1974b; Lheureux, 1976; Umanski et al., 1951). Hence despite the fact that in each case the number of non-irradiated connective tissue and epidermal cells transplanted was about the same, only skin cuffs which possessed a complete circumference could support normal distal transformation.

Additional experiments by Carlson (1975b) and Lheureux (1976) in which the epidermal and dermal components of the skin were separated in cuff transplantations, have shown that it is the dermal connective tissue layer which contains positional information, and that the epidermis is a passive partner in terms of pattern formation. Similar conclusions have been reached by Bryant and Tank (unpublished) from preliminary experiments on double-half limbs in axolotls and newts where restoration of a "complete circumference" in the epidermis is not sufficient to cause these limbs to undergo distal transformation.

Further information about those limb tissues that contain positional informations can be deduced from experiments in which a given tissue is placed into an abnormal circumferential location with respect to the remaining tissues of the stump. When cells with different circumferential positional values are grafted next to each other in this way, intercalation is expected to occur between them, leading to the formation of redundant sequences in the originally complete but reorganized circle of positional values. Carlson (1974b, 1975a, 1975b) and Lheureux (1972, 1975a) showed that when dermal connective tissue or muscle (which is covered by connective tissue) are transplanted to new circumferential positions, and amputation is performed through the grafted region, bizzare, multiple limbs are produced. Such regenerates can be considered to arise from a more than complete circle, i.e., one that contains redundant sequences, at the site of amputation, and they imply that connective tissue, at least, contains positional information. There are no experiments yet reported which would show whether or not the myocyte population of the muscles also contains positional information.

However, positional relocation of either the epidermis or the bone within the limb stump does not lead to the formation of multiple regenerates after amputation, so it may be concluded that these tissues do not contain positional information.

In summary, the results described in this section support the idea that positional information may be a property of a single tissue layer in the amphibian limb, the connective tissue. It is suggested that the epidermis on the outside, the myocytes within, and the bone or cartilage to the center of the regenerating limb acquire information about the pattern secondarily from the connective tissue layer.

V. CELL COMMITMENT DURING PATTERN REGULATION

In the previous sections, the picture we have developed of pattern regulation in amphibian limbs is one in which the type of new structures formed depends upon the the positional values of cells which are brought together by grafting or during migration or healing. There has been no suggestion so far that the types of new structures formed depend in any way upon the ancestry or lineage of those cells. In this section, we will examine the ways in which heritable commitments may in fact influence the types of new structures formed during growth and pattern regulation.

The first stable and heritable commitment which cells acquire during development is to a particular germ layer. Once the embryo has become subdivided into ectodermal, mesodermal and endodermal populations of cells, the progeny of these cells generally retain this commitment throughout all subsequent divisions. In the case of the vertebrate limb, which consists of ectoderm and mesoderm, these two cell types remain as separate lineages even during pattern regulation. When epidermis labelled with tritiated thymidine was grafted to an unlabelled newt limb stump (Riddiford, 1960) epidermal cells did not appear to contribute to the mesodermal blastema during regeneration. When the mesoderm was labelled and the epidermis unlabelled, mesodermal cells were found in the epidermis. This observation does not indicate that mesodermal cells had become epidermal cells, but rather that they had been incorporated into the epidermis by phagocytosis, during the normal process of wound cleaning (Singer and Salpeter, 1961). In addition Namenwirth (1974) has shown that when X-irradiated limb stumps are provided with an unirradiated graft of epidermis, no outgrowth occurs and no evidence can be found for epidermal cells giving rise to any other cell type.

A heritable difference arises between cells of the mesoderm itself when secondary fields are established. After that point in development mesodermal cells from the forelimb field give rise to cells which have qualities of the forelimb field and not the qualities of any other field, such as for example, that of the hind limb. This point has not been well documented in amphibians, since in these animals fore and hind limb can only be clearly distinguished by the number of digits formed (the forelimb has four digits, the hind limb five digits). The differences between tails and limbs are more distinctive and from various types of transplantation studies, it can be concluded that the mesodermal cells of these appendages carry a heritable commitment to field type (see Goss, 1961). In birds, there are clear differences between wings and legs, and when cells from the developing hind limb bud are transplanted to the developing wing bud the structures made by these cells and their progeny are characteristic of hind limb structures (Saunders *et al.*, 1957). This type of result also nicely illustrates the point that both heritable commitments and positional information play a role in the eventual fate of a group of cells, for if the hind limb cells are taken from the region of the prospective thigh, and grafted to the wing bud in the region of the prospective wing tip, then although the cells will form hind limb structures, these structures will be distal and not proximal structures.

A third type of possible heritable commitment in limb cells is to a particular tissue type within the mesoderm, i.e., to connective tissue, muscle, or bone and cartilage. The question of whether or not cells of one tissue type can give rise during distal transformation to those of another tissue type has interested students of regeneration for many years, but unambiguous answers have been slow to emerge. Steen (1968, 1970) showed that when cartilage cells, identifiable by both a tritiated thymidine label and triploidy, were grafted into the unlabelled diploid limb stumps of axolotls and allowed to participate in distal transformation, 96% of all the labelled cells in the regenerate were found in the new cartilage. However, when muscle, which contains connective tissue cells, myocytes and probably satellite cells or pericytes (Popiela, 1976) was the grafted tissue, labelled cells were found in all of the mesodermal tissues of the limb. These results indicate that cartilage cells have a stable heritable commitment to form cartilage, whereas at least some component of the muscle is able to undergo metaplasia. However, it was not clear to what extent the position of the graft in the stump affected the results, nor were these experiments designed to put any stress on the implanted cells to show their full potential for metaplasia. Recent experiments by Namenwirth (1974) and Dunis and Namenwirth (1977) have provided some more clear-cut results. In these experiments, axolotl limbs were X-irradiated to inhibit distal transformation, and then unirradiated tissue of one type or another was grafted into the limbs. The grafted cells and their progeny were marked by triploidy, whereas the host was diploid. All types of mesodermal tissue grafts were found to cause some outgrowth on the irradiated limbs, but not all were equally effective in permitting good morphogenesis. As might be expected if positional information resides in the cells of the connective tissue, grafts of skin more frequently led to the formation of normal distal regenerates than did grafts of cartilage. However, all mesodermal tissue grafts yielded some outgrowths which could be analyzed for the spectrum of differentiated cell types they contained. When this was done, it was found that some cell types can give rise to progeny which differentiate into other cell types. For example, connective tissue cells can form both connective tissue and cartilage in the outgrowth, and cartilage cells can form both cartilage and connective tissue. Muscle, which as mentioned above is a mixture of cell types, can give rise to all of the cell types of muscle and can also give rise to cartilage. It has not yet been technically possible to investigate the abilities of myocytes and satellite cells or pericytes separately from those of connective tissue cells, so the possibility still remains that the myocyte population of the muscle, which can only be formed from muscle, may in fact remain clonally distinct during distal transformation. Recent evidence from work on the chick limb suggests that in this vertebrate myocytes may have a separate origin from the remainder of the limb mesoderm during development (Chevallier *et al.*, 1977), but similar studies have not been carried out on amphibian limbs.

Finally, nothing is at present known about the interesting possibility that within a single tissue type in amphibian limbs there are different heritable commitments, analogous to those of the compartments in the epidermis of insects (Morata and Lawrence, 1977). The only hint of this possibility comes from studies of the patterns of innervation of the hind limb of salamanders (Diamond *et al.*, 1976). Here, nerves 15 and 17 occupy separate territories in the limb, and the boundary between the territories divides the limb into anterior and posterior fields of innervation. Further studies are necesary to investigate the possibility that anterior and posterior cells of a given tissue type may have different properties which are also inherited.

VI. CONCLUSIONS

During development or during pattern regulation, each cell eventually expresses a single differentiated phenotype. There appear to be two major types of information which determine the final decision of a cell to adopt one rather than any other differentiated state. The first is information which is stable and inherited, and which the cell shares with its relatives in the same lineage. The second is the information which the cell acquires according to its position within the field. The processes by which cells acquire these two different types of input have been termed determination and specification, respectively (P. Bryant, 1974).

In this article we have seen that during pattern regulation in amphibian limbs, local cell interactions leading to intercalation of new cells with appropriate intervening positional values provides a satisfactory formal explanation for a variety of experimental results. It has been suggested that the connective tissue cells of the limb carry stable positional information, and that the other cell types of the limb acquire information about the pattern secondarily. Finally, the extent to which various types of heritable commitment affect the final outcome during pattern regulation has also been considered. Experiments to test the existence of certain types of commitment have yet to be carried out in amphibians, and at present, we have no knowledge of the existence of compartments like those in insects, in any vertebrate.

ACKNOWLEDGMENTS

The author would like to thank Dr. Ian Sussex and the Society for Developmental Biology for the invitation to present this paper at the 36th Symposium. She would also like to thank Peter Bryant and Patrick Tank for critical comments on the manuscript; Warren Fox for expert technical assistance; and Lloyd Lemke for the art work. The research was supported by Grant Number HD 06082, awarded by the National Institutes of Health, DHEW.

REFERENCES

Brunst, V. V. (1950). *Quart. Rev. Biol.* **25**, 1-29.

Bryant, P. J. (1974). *Curr. Topics Develop. Biol.* **8**, 41-80.

Bryant, P. J., Bryant, S. V. and French, V. (1977). *Sci. Amer.* **237**, No. 1, 66-81.

Bryant, S. V. (1976). *Nature* **263**, 676-679.

Bryant, S. V. (1977). In: "Vertebrate Limb and Somite Morphogenesis" (D. A. Ede, J. R. Hinchliffe, and M. Balls, eds.), Cambridge University Press, Cambridge.

Bryant, S. V. and Baca, B. (1978). *J. Exp. Zool.* (in press).

Bryant, S. V. and Iten, L. E. (1976). *Develop. Biol.* **50**, 212-234.

Bryant, S. V. and Iten, L. E. (1977). *J. Exp. Zool.* **202**, 1-16.

Butler, E. G. (1955). *J. Morphol.* **96**, 265-281.

Carlson, B. M. (1974a). In: "Neoplasia and Cell Differentiation" (G.V. Sherbet, ed.), S. Karger, Basel.

Carlson, B. M. (1974b). *Develop. Biol.* **39**, 263-285.

Carlson, B. M. (1975a). *Develop. Biol.* **45**, 203-208.

Carlson, B. M. (1975b). *Develop. Biol.* **47**, 269-291.

Carlson, B. M., Civiletto, S. E. and Goshgarian, H. G. (1974). *Develop. Biol.* **37**, 248-262.

Chevallier, A., Kieny, M., Mauger, A., and Sengel, P. (1977). In: "Vertebrate Limb and Somite Morphogenesis (D. A. Ede, J. R. Hinchliffe, and M. Balls, eds.), Cambridge University Press, Cambridge.

Deck, J. D. (1955). *J. Morphol.* **96**, 301-331.

Dent, J. N. (1954). *Anat. Rec.* **118**, 841-856.

Diamond, J., Cooper, E., Turner, C. and Macintyre, L. (1976). *Science* **193**, 371-377.

Dunis, D. A. and Namenwirth, M. (1977). *Develop. Biol.* **56**, 97-109.

Faber, J. (1976). *Acta Biotheoret.* **25**, 44-65.

French, V., Bryant, P. J., and Bryant, S. V. (1976). *Science* **193**, 969-981.

Goss, R. J. (1961). *Adv. Morphogen.* **1**, 103-152.

Iten, L. E. and Bryant, S. V. (1975). *Develop. Biol.* **44**, 119-147.

Lheureux, E. (1972). *Ann. d'Embryol. Morphol.* **5**, 165-178.

Lheureux, E. (1975a). *Wilhelm Roux Arch.* **176**, 285-301.

Lheureux, E. (1975b). *Wihelm Roux Arch.* **176**, 303-327.

Lheureux, E. (1976). *Bull. Soc. Zool. Fr.* **101** (Suppl. No. 3), 109-118.

Maden, M. and Wallace, H. (1976). *J. Exp. Zool.* **197**, 105-114.

Maden, M. (1977). *J. Theor. Biol.* **69**, 735-753.

Morata, G. and Lawrence, P. A. (1977). *Nature* **265**, 211-216.

Morgan, T. H. (1901). "Regeneration." The MacMillan Company, New York.

Namenwirth, M. (1974). *Develop. Biol.* **41**, 42-56.

Popiela, H. (1976). *J. Exp. Zool.* **198**, 57-64.

Riddiford, L. M. (1960). *J. Exp. Zool.* **144**, 25-30.

Rose, S. M. (1962). In: "Regeneration" (D. Rudnick, ed.), Vol. 20, pp. 153-176. *Symp. Soc. Study Develop. Growth.*

Saunders, J. W. Jr., Cairns, J. M. and Gasseling, M. T. (1957). *J. Morphol.* **101**, 57-87.

Singer, M. (1974). *Ann. N. Y. Acad. Sci.* **228**, 308-322.

Singer, M. and Salpeter, M. M. (1961). In: "Growth in Living Systems." Basic Books, Inc., New York.

Smith, A. R. and Crawley, A. M. (1977). *J. Embryol. Exp. Morphol.* **37**, 33-48.

Smith, A. R., Lewis, J. H., Crawley, A., and Wolpert, L. (1974). *J. Embryol. Exp. Morphol.* **32**, 375-390.

Steen, T. P. (1968). *J. Exp. Zool.* **167**, 49-78.

Steen, T. P. (1970). *Am. Zool.* **10,** 119-132.

Stocum, D. L. (1975a). *Differentiation* **3,** 167-182.

Stocum, D. L. (1975b). *Develop. Biol.* **45,** 112-136.

Stocum, D. L. (1977). In: "Vertebrate Limb and Somite Morphogenesis" (D. A. Ede, J. R. Hinchliffe, and M. Balls, eds.), Cambridge University Press, Cambridge.

Summerbell, D. and Wolpert, L. (1972). *Nature* **239,** 24-26.

Summerbell, D. Lewis, J. H. and Wolpert, L. (1973). *Nature* **244,** 492-496.

Tank, P.W. (1978a). *Develop. Biol.* **62,** 143-161.

Tank, P.W. (1978b). *J. Exp. Zool.* (in press).

Tassava, R. A. and Mescher, A. L. (1975). *Differentiation* **4,** 23-24.

Thornton, C. S. (1968). *Adv. Morphogen.* **7,** 205-249.

Umansky, E. E., Tkach, V. K. and Kudokotsev, V. P. (1951). *Dokl. Akad. Nauk. SSSR* **76,** 465-467.

Mosaicism in the Central Nervous System of Mouse Chimeras

Richard J. Mullen

Departments of Neuropathology
Harvard Medical School and Neuroscience
Children's Hospital Medical Center
Boston, Massachusetts 02115

I. INTRODUCTION

When a cell divides, it gives rise to two daughter cells. For at least that brief moment following cell division there exists a clone of two cells. Thus, one question we are addressing in this Symposium is not so much whether or not clones exist but, rather, whether or not these clones are maintained. Do daughter cells remain contiguous and give rise to larger and larger clones or are the movements and migrations of cells during development sufficient to disrupt the clones? The larger question to be asked is of what significance is the presence or absence of clones in the nervous system of the developing and mature organism. There are numerous anatomical and physiological properties of the CNS that might be explained in terms of clones or at least better understood if we knew more about cell lineages.

If the disruption of clones by migration were complete, that would give rise to a random distribution of cells. It would be difficult to believe that the end result of the magnificent process of nervous system development that we study is a random distribution of cells. Nevertheless, in the developing mammalian central nervous system, cell migration is, in fact, more the rule than the exception. Therefore, before discussing the CNS of mouse chimeras, I will give some examples of cell movement in the developing CNS with particular attention to whether the cell movements are likely to disrupt or maintain clones.

When one looks at the CNS, the presence of layers of cells of different types in both retina and the cortices of the brain is striking. These different cell types arise at different times during development, but most come from the same proliferative zone. Do these different cell types in a given region come from one population of precursors or are there several populations? If they do come from common ancestors, it opens up some intriguing possibilities. When the cortex is stimulated, it is not different layers of cells that respond, but rather columns of cells that include all cell layers. This interconnectivity could be determined relatively late in development or be the result of early contacts made between cells during radial migration. However, it might also be a result of the columns being clonally related. In other words, the specificity of some synapses might be controlled by cells synapsing on relatives or ancestors. This hypothesis has been put forth previously by Miale and Sidman (1961) and Jacobson (1968).

In regions of brain below the cortex there are numerous relatively small populations of cells called nuclei. Are these nuclei clones of neurons? In the cerebellum and cerebrum it is possible to physiologically map out regions of cortex responding to and controlling different parts and functions of the body. Do these regions have a clonal basis to them? Most of these questions are enormously complex and beyond the scope of this paper. However, we can take the first step by taking some glimpses of mosaicism in the CNS, particularly in the cerebellum and retina. The final answers to these questions have significance not only for development, but also for nervous system function.

II. DEVELOPMENT OF THE MAMMALIAN CNS

In the studies on chimeras that follow we will be considering the positions and distribution of cells in adult animals. Since most cell types are not generated *in situ* but rather migrate to their respective positions during development, the degree to which clones are maintained or disrupted would be determined in part by the manner and patterns of cell migrations. Neuronal migration and cortical development is a fascinating but complex subject. Here, I will only attempt to give some simple generalizations that are particularly relevant to the topic of clones. More detailed descriptions may be found in reviews by Sidman and Rakic (1973) and Caviness and Rakic (1978).

Early in organogenesis the neural tube appears as a columnar epithelium, the cells being in contact with both inner and outer surfaces (Fig. 1A). During the process of cell division the nucleus moves back and forth from the inner ventricular surface. While it is away from the surface it synthesizes DNA but then returns to the ventricular surface where cell division occurs. This process is repeated over and over. The day of development on which the progeny of these ventricular cells have gone through their final cell division is said to be the "birthday" of those particular cells. Each neuronal cell type has its specific "birthdates". Cerebellar Purkinje cells in the mouse, for example, are all born on embryonic days 11 to 13 (Miale and Sidman, 1961). Once born, the young neurons migrate radially to take up their class-specific position in the developing cortex (Fig. 1B).

Most of the neurons generated subsequently do not take up positions below neurons born earlier, but rather migrate past them to occupy more superficial positions (Fig. 1C). This surprising phenomenon has come to be known as "inside-out" development (Angevine and Sidman, 1961). In some instances, this "inside-out" development results in later generated neurons having to migrate considerable distances through an increasingly complex cortex. Rakic (1971, 1972) has found that in some regions young neurons migrate along radial glial guides to reach these increasingly distant layers of cortex (Fig. 1D).

In the foregoing discussion, summarized diagrammatically in Fig. 1A-D, I believe the key word is "radial". From the simple columnar epithelium to the layering of cells in cortex, the orientation and migration of cells is in a radial direction. Because of this common orientation, it appears it would be possible for successive generations of daughter cells to remain relatively contiguous. Thus, despite migration, the maintenance of clones appears possible.

The final diagram (Fig. 1E) depicts one of the most remarkable migration events in the CNS and also serves to introduce some of the cell types and architecture of the cerebellum. Early in development, a population of cells that are still mitotic leaves the ventricular zone to form a subventricular zone. From this zone in the rhombic lip arises a population of cells whose progeny are destined to become the cerebellar

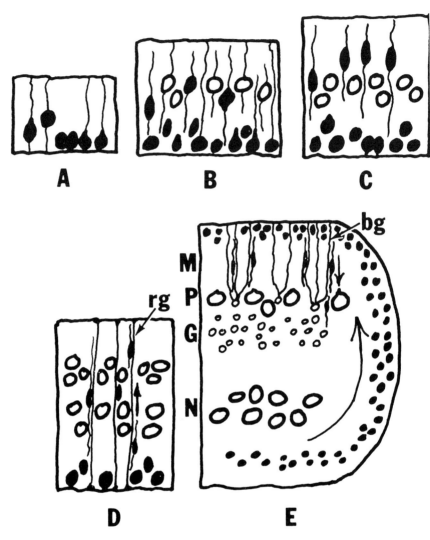

Fig. 1. Schematic diagrams of some events in CNS development. In each, the ventricular surface is at the bottom, the pial surface at the top. Filled cells represent cells that are either still capable of dividing or are in the process of migration. Open cells represent cells that have obtained the relative position they will occupy in the mature CNS. (A) Columnar epithelium of neural tube; to and fro movement of nuclei. (B) Radial migration of neurons following last cell division. (C) "Inside-out" development: cells migrating past cells born earlier. (D) Migration of cells along radial glia (rg). (E) Cerebellar development: external granule cells (small filled cells) migrate over the surface of the cerebellum (large arrow), then migrate inwards (small arrow), along the Bergmann glial fibers (bg) of the Golgi epithelial cells. Mature cerebellum comprised of molecular layer (M), Purkinje cell layer (P), granule cell layer (G), and the deep nuclei (N). See text for further explanation. Adapted from drawings by Rakic (e.g., in Sidman and Rakic, 1973).

granule cells. However, rather than migrating radially to the developing granule cell layer that lies below the Purkinje cell layer, the population migrates over the surface of the cerebellum. This migrating population, which is still mitotic, is the transitory external granule cell layer. To reach their final position, the granule cells migrate inward, past the Purkinje cells, leaving behind them their axons which synapse on Purkinje cell dendrites in the molecular layer. The granule cells are guided through the molecular layer by the Bergmann glial fibers of the Golgi epithelial cells (Rakic, 1971).

The Purkinje cells, which are the only cells to send their axons out of the cortex, synapse on cells in the deep cerebellar nuclei. Both of these cell types arose at about the same time from the ventricular zone (Miale and Sidman, 1961; Taber-Pierce, 1975). The mature cerebellum thus consists of a cortex and the deep nuclei. The cortex is comprised of three layers: the cell sparse molecular layer, the Purkinje cell layer and the granule cell layer. The cortex folds repeatedly, giving rise to the leaflike folia, the larger of which are called lobules.

III. METHODS FOR OBSERVING MOSAICISM IN THE CNS

One of the major problems in studying the CNS of mouse chimeras is being able to identify the genotype of individual cells. At present, there is no single technique available that can be used for all cell types. The techniques that are available fall into two categories. The first is the use of mutant genes that destroy or markedly alter particular cell types. The second is the use of variations, such as enzyme levels, between different strains of normal mice as cell markers.

A. *Neurological Mutants*

There are several well-characterized neurological mutants in mice that lose a particular cell type. To use these as markers for the particular cell type, however, it is necessary to demonstrate that the mutant gene acts intrinsically. That is, it must be demonstrated that the mutant cells degenerate and the genotypically normal cells survive. Studies aimed at determining the site of gene action have been recently reviewed elsewhere (Mullen, 1977a) and will only briefly be described here.

Purkinje cell degeneration (*pcd*) is a mutant that postnatally loses virtually all of its cerebellar Purkinje cells (Mullen *et al.*, 1976). In *pcd/pcd<—>+/+* chimeras a mosaic of surviving cells and spaces left by degenerated cells is observed. With β-glucuronidase (described below and in Fig. 2A) as a cell marker, it was found that the surviving cells were all genotypically +/+ (Fig. 2B); the *pcd/pcd* cells had all degenerated (Mullen, 1977a, b). Thus, the *pcd* locus acts within the Purkinje cell and can be used to study the distribution of the surviving normal cells in *pcd/pcd<—>+/+* chimeras.

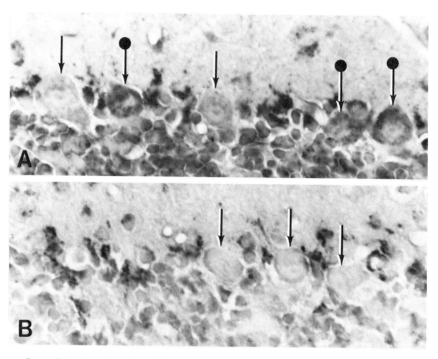

Fig. 2. Sagittal sections of cerebellar cortex stained histochemically for β-glucuronidase (the red precipitate appears dark gray in these micrographs) and counterstained with methyl green. (A) Section from a $Gus^b/Gus^b<—>Gus^h/Gus^h$ chimera. Some of the Purkinje cells are stained (arrows with dots) and are, therefore, Gus^b/Gus^b; others are not stained (arrows) and are, therefore, Gus^h/Gus^h. (B) Section from a $Gus^h/Gus^hpcd/pcd<—>Gus^h/Gus^h+/+$ chimera. Since none of the surviving Purkinje cells (arrows) are stained, they are from the $Gus^h/Gus^h+/+$ component. (From Mullen, 1977b).

Staggerer (*sg*) mutants lose cerebellar granule cells (Sidman *et al.*, 1962). However, it does not appear it will be useful as a marker. Our analysis of *sg/sg<—>+/+* chimeras indicates the gene is acting in Purkinje cells (Herrup and Mullen, 1976) as was suggested by earlier electron microscopic studies (Landis, 1971; Sidman, 1972; Hirano and Dembitzer, 1975). We have not yet determined whether the gene also acts in granule cells.

In reeler (*rl*) mutants there is no cell loss but most cortical neuronal cell types are out of position. This mutant also cannot be used as a marker because we have found the cells are not being positioned according to their own genotype. In *rl/rl<—>+/+* chimeras we found that patches of aberrantly positioned cells contained both *rl/rl* and *+/+* cells (Mullen, 1977a; Mullen and Sidman, in preparation).

Retinal degeneration (*rd*) causes degeneration of photoreceptor cells and has been used in several studies of chimeras (Mintz and Sanyal, 1970; Wegmann *et al.*, 1971; LaVail and Mullen, 1976a, b; Sanyal and Zeilmaker, 1976; West, 1976a). From the studies of mutants and chimeras (reviewed in LaVail and Mullen, 1976b) the site of gene action has been localized to the neural retina. However, it must be noted that because of lack of a cell marker it is not definitively known to be acting in the photoreceptor cell although I know of no reason to think it is not.

B. *Cell Markers*

Although there are approximately fifty enzyme variants known in the mouse, most are electrophoretic variants and cannot be used as individual cell markers. The structural loci for β-glucuronidase (*Gus*) is one exception and was first used histochemically by Condamine *et al.* (1971) to demonstrate mosaicism in the livers of chimeric mice. Some strains of mice carry the Gus^b allele and have "high" or "normal" enzyme activity, whereas other strains carry the Gus^h allele and have "low" enzyme activity in all tissues. The histochemical procedure utilizes pararosanilin so that a red precipitate is deposited at sites of enzyme activity. Recently, Feder (1976) has developed a more sensitive histochemical procedure that makes it possible to use the enzyme as a cell marker on tissues other than liver. We have found that in the nervous system it can be used as a marker at least for Purkinje cells (Fig. 2A), choroid plexus, some nuclei in the brain stem and perhaps a few other cell types. It cannot be used on all cell types because of uptake of enzyme by low activity, Gus^h, cells. Feder (1976) observed increased staining in low cells of liver and other tissues and interpreted it as resulting from uptake of enzyme. To distinguish between uptake of enzyme and induction or increased synthesis of Gus^h enzyme we took advantage of the decreased thermostability of the Gus^h enzyme. When sections of liver from $Gus^b/Gus^b<—>Gus^h/Gus^h$ chimeric mice were heat-treated under conditions that destroyed virtually all enzyme activity in Gus^h/Gus^h controls, the low cells in the chimeras still exhibited enzyme activity (Herrup *et al.*, 1976; Herrup and Mullen, 1978). We interpreted this heat stable enzyme as being Gus^b enzyme that was taken up by Gus^h cells.

In view of our studies on uptake of enzyme, can β-glucuronidase be used as a Purkinje cell marker? We have used the technique on dozens of chimeras and are convinced of its validity. In fact, Purkinje cells appear to be the only cells in the brain that do not take up the enzyme. In our preparations of $Gus^b/Gus^b\!<\!-\!>\!Gus^h/Gus^h$ chimeras, virtually every Purkinje cell can be classified as being either stained or not stained. More conclusive evidence of its validity comes from studies on mutant chimeras. In Gus^b/Gus^b $pcd/pcd\!<\!-\!>\!Gus^h/Gus^h$ $+/+$ chimeras none of surviving Purkinje cells were stained (Fig. 2B). Also, in Gus^b/Gus^b $sg/sg\!<\!-\!>\!Gus^h/Gus^h$ $+/+$ chimeras, Purkinje cells that were normal in appearance and position (i.e., the $+/+$ cells) were not stained, whereas those that were abnormal (i.e., sg/sg) were stained (Herrup and Mullen, 1976; Mullen, 1977a). These studies, plus the fact that the fine grained mosaicism of Purkinje cells we observe in β-glucuronidase chimeras (described below) has also been observed by Dewey *et al.* (1976) using other cell marking techniques, attest to the validity of this marker.

Dewey *et al.* (1976) have visualized Purkinje cell mosaicism using chimeras whose cells differed in levels of the β-galactosidase *(Bgs)* enzyme. They also used an immunofluorescence technique directed against histocompatibility-antigen differences.

IV. MOSAICISM IN THE BRAIN

A. *Purkinje Cells*

We first observed mosaicism in the brain in sagittal sections of cerebellum of a Purkinje cell degeneration chimera (i.e., $pcd/pcd\!<\!-\!>\!+/+$) (Mullen, 1975). In single sections it was obvious that many Purkinje cells had degenerated and the remaining ones appeared scattered along the Purkinje cell layer. The surviving cells appeared singly or in small groups. The impression given by single sections, however, could be misleading. The single cells, for example, could result from the plane of section passing through the edge of a large patch or row of cells. To investigate this possibility a serial reconstruction of part of a single cerebellar lobule was undertaken.

The lobulus centralis (Atlas of Sidman *et al.,* 1971) was selected for reconstruction because of its relatively simple shape and moderate size (only one movement of the microscope slide was needed to visualize the entire lobule). Based on comparison of cell counts with normal animals, approximately 75% of the cells in this lobule had degenerated. Using a drawing tube, the positions of the Purkinje cells were marked on tracing paper. After superimposing the tracings to align them, the positions of cells along the lamina were measured and then plotted in a straight line on graph paper. The reconstruction is shown in Fig. 3. It is an enlarged version (i.e., more sections) of one presented earlier (Mullen, 1977a). There does not appear to be any coherent patches or clones other than what one would expect in a random distribution.

Although the Purkinje cell layer is only one cell thick, it is a curved surface and so when reconstructed as a flat, two-dimensional array there will be some distortion. However, it should be emphasized that although this technique of reconstruction might distort the shape of a patch of cells, it would not disrupt any patches.

The distribution of cells shown in Fig. 3 does not result from mixing of cells at an interface of a completely normal and a completely degenerated patch. Regions of cerebellum surrounding the region reconstructed were examined and there was no region where the Purkinje cells were present in normal numbers nor completely degenerated.

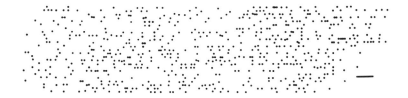

Fig. 3. Serial reconstruction (twenty-five 20 μm sections) of part of the lobulus centralis of a pcd/pcd<—>+/+ chimera in which 75% of the Purkinje cells had degenerated. Size of the dots is approximately equal to the mean diameter of the Purkinje cells (14 μm). The dots were plotted in a straight line (as if the cells were in the middle of the section) unless the cell was present on an adjacent section. In the latter case, the dot was plotted slightly towards the adjacent section. Bar represents 100 μm. (From Mullen, 1977b).

A more difficult question is whether there was any rearrangement of +/+ cells after degeneration of the *pcd/pcd* cells. In *pcd/pcd* mutants, the degeneration occurs relatively late, between postnatal days 20 and 40. During this period the cerebellum is almost fully matured. It is difficult to visualize Purkinje cells with their massive dendrites that extend across the molecular layer and receive hundreds or thousands of synaptic contacts rearranging themselves. It is nevertheless a possibility and can be investigated by examining mosaicism in chimeras that have not lost any cells.

In single sections of cerebellum from chimeras in which there was no cell loss, the mosaicism observed with cell markers appears as fine grained as that observed in *pcd* chimeras. In our laboratory, we observed this in β-glucuronidase chimeras (Gus^b/Gus^b<->Gus^h/Gus^h) where some Purkinje cells were stained red (i.e., Gus^b) whereas others were not stained (i.e., Gus^h) (Mullen, 1977a and Fig. 2A). Dewey *et al.* (1976) had made a similar observation using the β-galactosidase and immunofluorescent techniques. To do a serial reconstruction of sections from a β-glucuronidase chimera would be extremely difficult. Fot Purkinje cells, the histochemical procedure is best done on 7-8 μm sections. Since the mean diameter of the Purkinje cells in these preparations is nearly twice as great (approximately 14 μm), virtually every cell would be present on at least two, and sometimes three, adjacent sections. The glucuronidase chimeras, therefore, have been analyzed by two other methods. The first involved estimating the number of cells in a clone from data on the mean number of cells in a patch. The second involved determining the proportion of cells of like genotype in different regions.

It is important to distinguish between a "patch" and a "clone". As clearly stated by Nesbitt (1974) "clone will designate a group of clonally related cells which have remained contiguous throughout the history of the embryo" and "patch will designate a group of cells of like genotype which are contiguous at the moment of consideration". The distinction is important because in a completely randomized population some cells of the same genotype would be contiguous. The frequency of these contiguous cells would depend on the frequency of that genotype in the population; the higher the frequency, the more contiguous cells and hence the larger the mean patch size. As shown by Roach (1968) and discussed in relation to chimeras by West (1975), the mean patch size in a random linear array can be estimated by the simple formula $1/(1-p)$, where p is the proportion of cells of that genotype in the population. The number of cells in a clone can be estimated by dividing the observed mean patch size by the expected.

The "railroad track" shown in Fig. 4 is a tracing of the entire Purkinje cell layer in a section from a $Gus^b/Gus^b<->Gus^h/Gus^h$ chimera. It shows the positions of "red" Gus^b cells and unstained Gus^h cells. To estimate the size of a clone of Purkinje cells, five sections, such as that in Fig. 4, were analyzed. Alternating sections were used so that the same cell would not be counted twice. The results are shown in Table I. In the sections analyzed, 17.5% of the 1947 cells were Gus^b. The observed mean patch size was 1.259 cells. In a completely random array the estimated patch size would be 1.212. Thus, in this chimera, a "clone" would be comprised of 1.03 Purkinje cells*. This data suggests that the Purkinje are nearly randomly distributed. It also is evidence that the distribution of surviving cells shown in the serial reconstruction of the *pcd* chimera in Fig. 3 is not the result of cell rearrangement following degeneration.

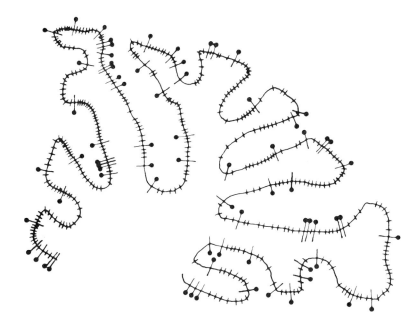

Fig. 4. Tracing of the Purkinje cell layer of a $Gus^h/Gus^b<->Gus^h/Gus^h$ chimera showing the positions of unstained Gus^h cells (cross marks) and stained Gus^b cells (large cross marks with dots). Sagittal section near the midline of the cerebellum.

*These figures are based on a linear array. For an approximation of a two-dimensional array the numbers could be squared. That would not, however, change the interpretation.

TABLE I.

Mean Patch Size and Estimate of Clone Size for Gus^b Purkinje
Cells in a Gus^b<—>Gus^h Chimera*

Mean Number of Gus^b Cells per Patch (5 sections)				
1.245±0.071	1.367±0.100	1.196±0.051	1.298±0.091	1.217±0.063

	Cells
Observed Mean Patch Size (Total)	1.259±0.033
Expected in Random Linear Array [1/(1-pred)]	1.212
Clone (observed/expected)	1.030±0.021

*Based on 1947 cells; 270 patches; 340 "red" Gus^b cells; [pred]=0.175.

Although the above analysis suggested a near random arrangement of cells, the raw data appeared to show slight variation in the proportions of the two cell types in different cerebellar regions. To pursue this, nine sagittal sections of the cerebellar vermis (near the midline) were divided into three regions: the anterior, middle and posterior lobules. Approximately 3500 cells were recorded and the proportion of "red" Gus^b cells in each of the three regions was calculated. The results are shown in Table II. There was no significant difference between the proportion of Gus^b cells in the anterior and middle regions, but there was a significantly higher proportion in the posterior region (14-15% versus 26%).

TABLE II.

Proportion of Gus^b Cells [pred] in Three Regions of the
Cerebellar Vermis of a Gus^b<—>Gus^h Chimera*

Anterior	Middle	Posterior
0.141±0.006	0.150±0.011	0.263±0.015**

*Means from 9 sagittal sections; 3525 cells.
**Significantly greater, P<0.01.

Thus, within the Purkinje cell layer of chimeras there are regional variations in the proportions of the two cell types which probably reflect earlier clonal development. Within a given region in the mature cerebellum, however, the two populations are almost, if not actually, randomly distributed. This indicates that during some specific period of development or throughout development of the cerebellum there was an enormous amount of cell mixing.

B. *Other Cell Types in the Brain*

In *Gus^b/Gus^b* mice, the choroid plexus is a tissue rich in β-glucuronidase and in *Gus^b/Gus^b<—>Gus^h/Gus^h* the mosaicism is obvious. We have not analyzed the mosaicism but it appears that the two cell types, though probably not randomly distributed, are present in small patches (Fig. 5).

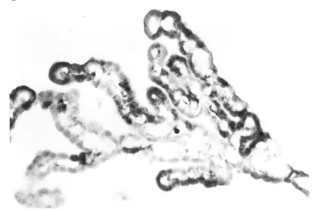

Fig. 5. Mosaicism in a section of the choroid plexus of a *Gus^b/Gus^b<—>Gus^h/Gus^h* chimera stained for β-glucuronidase, counterstained with methyl green. The *Gus^b* cells show intense staining while the *Gus^h* cells show relatively little staining.

For other cell types in the CNS, the β-glucuronidase marker must be used with caution because of the problem of uptake of enzyme discussed above. For example, in the cerebral cortex, which is an area that we are most anxious to study, there are some cells in chimeras that are intensely stained. These cells are undoubtedly *Gus^b*. There are, however, too many cells that are ambiguous. In other cell types, such as some of the large neurons in brain stem nuclei, mosaicism is evident despite the increased staining in the "low" *Gus^h* cells. An example is the hypoglossal nucleus shown in Fig. 6. Thus, this nucleus is not a clone and, at least in single sections, there does not appear to be any pattern to the positions of the two cell populations.

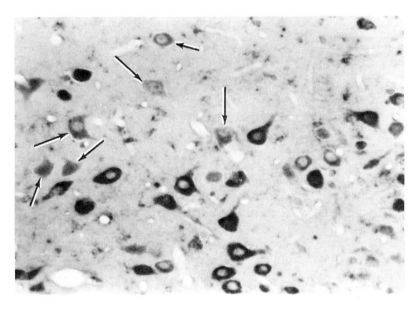

Fig. 6. Section through the hypoglossal nucleus of the medulla of a β-glucuronidase chimera. There are two populations of cells; intensely stained Gus^b/Gus^b cells and lightly stained Gus^h/Gus^h cells (arrows). Counterstained with methyl green.

V. MOSAICISM IN THE RETINA

The area of the central nervous system of chimeras that has received the most attention is the retina, specifically, the photoreceptor cells. The marker that has been used for photoreceptor cells is the retinal degeneration (rd) gene. As mentioned above in section IIIA, this locus is probably acting in the photoreceptor cell although, because of lack of an independent cell marker, it has not been conclusively demonstrated. In the following section, the assumption is made that, in chimeras, all rd/rd cells, and only rd/rd cells, degenerate.

Mosaicism in the pigment epithelium of pigmented<—>albino chimeras is, of course, easily observed by the presence or absence of pigment granules in the cells.

A. Photoreceptor Cells

As shown originally by Mintz and Sanyal (1970) and Wegmann *et al.* (1971), chimeras between strains homozygous for *rd*, retinal degeneration, and normal mice show a patchy degeneration in the retina. Based on their analysis of the distribution of degenerated regions, Mintz and Sanyal (1970; see also Mintz, 1971) concluded that the basic plan consisted of 10 sectors, or clones, radiating from the back of the eye. It has been noted that in chimeras there are regions of intermediate retina where the degeneration is not complete, nor is the retina normal (Mintz and Sanyal, 1970; Wegmann *et al.*, 1971; LaVail and Mullen, 1976a, b; West, 1976a; Sanyal and Zeilmaker, 1976).

Fig. 7 shows a two-dimensional reconstruction of a retina from a *rd/rd*<—>+/+ chimera (LaVail and Mullen, 1976a). There are obviously radiating sectors visible in this reconstruction. It is also obvious that regions of intermediate retina are very extensive. If we assume, as stated above, that all *rd/rd* cells and only *rd/rd*, degenerate, then how does

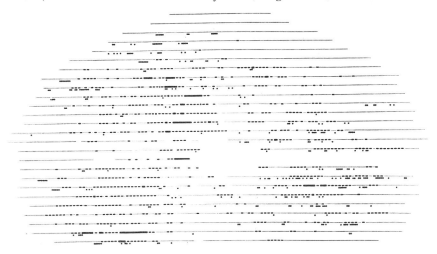

Fig. 7. Two-dimensional map of the retina and pigment epithelium of an *rd/rd/ c/c*<—>+/+ +/+ chimera. Reconstruction based on every tenth 10 μm section. Solid lines represent normal retina (5 or more rows of photoreceptor cell nuclei); dashed lines represent intermediate retina (1-4 rows of nuclei), and dotted lines represent completely degenerated retina. The small black squares below the section lines represent pigmented pigment epithelial cells (i.e., from the +/+ +/+ component). (From LaVail and Mullen, 1976a).

intermediate retina come about? One possibility, depicted in Fig. 8A, is that there is movement of +/+ photoreceptor cells after degeneration of the *rd/rd* cells (LaVail and Mullen, 1976a, b; West, 1976a). That this might be occurring is supported by the observations that there might also be cell movement in the inner nuclear layer (the layer beneath the photoreceptor cells) (West, 1976a).

The extensiveness of the intermediate patches, however, seems to warrant some other explanation. As depicted in Fig. 8B, intermediate patches would also result from mixing of cells during development, before degeneration of the *rd/rd* cells (LaVail and Mullen, 1976a, b). The extent to which these two processes contribute to the intermediate patches can only be assessed by examining mosaicism in retinas that have not lost any cells. Unfortunately, at present, there is no photoreceptor marker available to help answer this question.

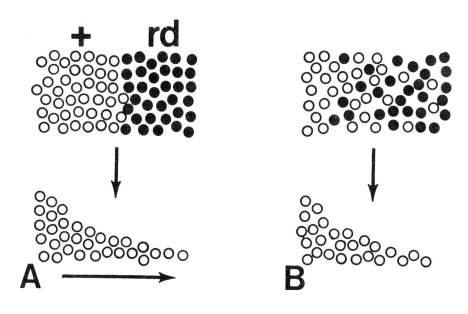

Fig. 8. Two possible mechanisms that could lead to the formation of regions with intermediate amounts of degeneration in the retinas of *rd/rd*<–>+/+ chimeras. (A) If there are rigid clones of +/+ cells (open circles) and *rd/rd* cells (filled circles) prior to degeneration, then intermediate retina must result from movement of +/+ cells (horizontal arrow) following degeneration. (B) The same intermediate patches could result from intermingling of *rd/rd* and +/+ cells before degeneration of the *rd/rd* cells.

If future studies indicate that the intermediate patches result primarily, or exclusively, from movement of cells after degeneration, it will mean that photoreceptor cell development is quite rigid in that clones are maintained. If, however, the movement of cells is minimal, it will mean there is extensive cell mixing during development.

B. *Pigment Epithelial Cells*

Mosaicism in the pigment epithelium of the retina is so easily observed, it is not surprising that it has been reported in numerous studies (Mystkowska and Tarkowski, 1968; Mintz and Sanyal, 1970; Deol and Whitten, 1972; Sanyal and Zeilmaker, 1976; LaVail and Mullen, 1976a, b). Radiating patches, evidence of clonal development, was first noted by Mintz and Sanyal (1970; see also Mintz, 1971). The most extensive analysis is in a recent report by Sanyal and Zeilmaker (1977). They presented computerized reconstructions of the distribution of pigmented and albino cells in the pigment epithelium of 20 eyes. It is obvious from their reconstructions that the distribution of cells is not random; there is obvious evidence of the radiating sectors reported by Mintz and Sanyal (1970). I think it is also obvious, however, that there is extensive cell mixing.

West (1976b) has used the analysis of patch size in sections to estimate the number of cells per coherent clone (described above in Section IVA). His analyses suggested that at embryonic day 12.5, the distribution of cells in the pigment epithelium was almost random while in adults, there were approximately six cells per coherent clone.

VI. SUMMARY, CONCLUSION AND SPECULATION

The title of this Symposium is The Clonal Basis of Development. We have been shown beautiful examples of clonal development in both plants and animals. The theme of my talk, on the contrary, has been the extensive cell mixing in the developing mammalian central nervous system.

It was pointed out in the beginning that, although cells move extensively in the developing CNS, the movement was primarily radial and it appeared there were mechanisms for maintaining clones. It seems puzzling that there are such sophisticated means for orderly moving cells in one dimension (i.e., radial) only to have the cells quite thoroughly mixed in the other dimensions. From studies of the reeler (*rl*) mutant mouse (e.g., Caviness, 1977), it is known that positioning of cells along the radial axis is critical to the formation of a normal CNS. It may be that if cells achieve their proper radial position, their subsequent deployment throughout their respective cell layers is not important. On the other hand, we might speculate that what appears to be a carefree randomization of, for example, Purkinje cells in the cerebellum may actually represent a new and equally important dimension of cell migration. If cell lineage is a determinant in the formation of some primary types of synapses, then this apparent randomization may actually be the result of an active and controlled migration of cells aimed at distributing daughter cells in different parts of the cortex. As a result, incoming information could be assimilated from a larger region.

Future studies on other brain regions, such as cerebral cortex, may reveal more of a clonal type of development. If they do not, however, it should not be looked upon as a disappointment but rather as an indication that our understanding of mammalian nervous system development is far from complete. A complete understanding of the development and genetics of the nervous system will undoubtedly be aided by studies of chimeras.

REFERENCES

Angevine, J. B., Jr. and Sidman, R. L. (1961). *Nature* **192**, 766-768.

Caviness, V. S., Jr. (1977). *In:* "Society for Neuroscience Symposia" (W. M. Cowan and J. A. Ferrendelli, eds.), Vol. II, pp. 27-46, Society for Neuroscience, Bethesda.

Caviness, V. S., Jr. and Rakic, P. (1978). *Ann. Rev. Neurosci.* 1:297-326.

Condamine, H., Custer, R. P. and Mintz, B. (1971). *Proc. Nat. Acad. Sci. U.S.* **68**, 2032-2036.

Deol, M. S. and Whitten, W. K. (1972). *Nature New Biol.* **238**, 159-160.

Dewey, M. J., Gervais, A. G. and Mintz, B. (1976). *Develop. Biol.* **50**, 68-81.

Feder, N. (1976). *Nature* **263**, 67-69.

Herrup, K. and Mullen, R. J. (1976). *Neurosci. Abstr.* **2**, 101.

Herrup, K. and Mullen, R. J. (1978). (submitted).

Herrup, K. and Mullen, R. J. and Feder, N. (1976). *Fed. Proc.* **35**, 1371.

Hirano, A. and Dembitzer, H. M. (1975). *J. Neuropath. Exp. Neurol.* **34**, 1-11.

Jacobson, M. (1961). *Develop. Biol.* **17**, 202-218.

Landis, D. (1971). *J. Cell Biol.* **51**, 159A.

LaVail, M. M. and Mullen, R. J. (1976a). *Exp. Eye Res.* **23**, 227-245.

LaVail, M. M. and Mullen, R. J. (1976b). In "Retinitis Pigmentosa" (M. B. Landers, M. L. Wolbarsht, J. E. Dowling and A. M. Laties, eds.), pp. 153-173. Plenum Press, New York.

Miale, I. L. and Sidman, R. L. (1961). *Exp. Neurol.* **4**, 277-296.

Mintz, B. (1971). *Fed. Proc.* **30**, 935-943.

Mintz, B. and Sanyal, S. (1970). *Genetics* (Suppl.) **64**, 43-44.

Mullen, R. J. (1975). *Genetics* (Suppl.) **80**, 56.

Mullen, R. J. (1977a). In: "Society for Neuroscience Symposia" (W. M. Cowan and J. A. Ferrendelli, eds.), Vol. II, pp. 47-65. Society for Neuroscience, Bethesda.

Mullen, R. J. (1977b). *Nature* 270: 245-247.

Mullen, R. J., Eicher, E. M. and Sidman, R. L. (1976). *Proc. Nat. Acad. Sci. U.S.* **73,** 208-212.

Mystkowska, E. T. and Tarkowski, A. K. (1968). *J. Embryol. Exp. Morphol.* **20,** 33-52.

Nesbitt, M. M. (1974). *Develop. Biol.* **38,** 202-207.

Rakic, P. (1971). *J. Comp. Neur.* **141,** 283-312.

Rakic, P. (1972). *J. Comp. Neur.* **145,** 61-84.

Roach, S. A. (1968). "The Theory of Random Clumping". Methuen, London.

Sanyal, S. and Zeilmaker, G. H. (1976). *J. Embryol. Exp. Morphol.* **36,** 425-430.

Sanyal, S. and Zeilmaker, G. H. (1977). *Nature* **265,** 731-733.

Sidman, R. L. (1972). In: "Cell Interactions: Proceedings of the Third Lepitit Colliquium" (L. G. Silvestri, ed.), pp. 1-12. North-Holland, Amsterdam.

Sidman, R. L. and Rakic, P. (1973). *Brain Res.* **62,** 1-35.

Sidman, R. L., Lane, P. and Dickie, M. (1962). *Science* **137,** 610.

Sidman, R. L., Angevine, J. B., Jr. and Pierce, E. T. (1971). "Atlas of the Mouse Brain and Spinal Cord". Harvard University Press, Cambridge.

Taber-Pierce, E. (1975). *Brain Res.* **95,** 503-518.

Wegmann, T. G., LaVail, M. M. and Sidman, R. L. (1971) *Nature* **230,** 333-334.

West, J. D. (1975). *J. Theor. Biol.* **50,** 153-160.

West, J. D. (1976a). *J. Embryol. Exp. Morphol.* **36,** 145-149.

West, J. D. (1976b). *J. Embryol. Exp. Morphol.* **35,** 445-461.

Clonal Analysis of Behavior in Mice

Muriel N. Nesbitt

Department of Biology
University of California, San Diego
La Jolla, California 92093

Karla Butler

Department of Psychology
California State University
Northridge, California 91324

M. Anne Spence

Department of Psychiatry
Mental Retardation Unit/NPI
University of California, Los Angeles
Log Angeles, California 90024

I. INTRODUCTION

Mice of the C57BL inbred strains differ from mice of the A strains with respect to a great many characteristics including a variety of behavior traits. Since chimeras of the C57<−>A constitution are mixtures of A and C57 cells, with varying proportions and distributions of the two cell types, they can be used to learn certain things about the natures of the differences between A and C57 mice. We have studied the behavioral characteristics of a series of A<−>C57 chimeras for the purpose of discovering (i) whether any of the behaviors in which A and C57 mice differ are controlled by single clones, and (ii) how many discrete, separately determined differences underlie the phenotypic differences we measure.

II. THE BEHAVIORS

We limited our study to four rather easily measured behavioral categories in which A and C57 mice differ dramatically: open field activity (McClearn, 1959, 1960), alcohol preference (McClearn, 1972; McClearn and Rodgers, 1961), cricket killing (Butler, 1973), and rope climbing.

A. Open Field Activity

The open field test was carried out by placing the mouse to be tested in a box with a meter-square floor area marked off with painted lines into a 64 square gridwork. Each mouse was placed in the center of the gridwork at time 0, and given 5 minutes in which to explore the box. In our hands C57BL/6J mice run across 206±69 squares during the test period, while A/J run across an average of 62±21. Another difference between these strains which can be scored during this test is the tendency to defecate in the open field. C57BL/6J mice rarely defecate during the test while A/J mice defecate 5±1 times.

B. Alcohol Preference

We tested alcohol preference by providing each mouse to be tested with two graduated drinking bottles, one containing water, and the other containing a solution of 10% ethanol in water. The amount of fluid consumed from each bottle was measured daily. Every fourth day the alcohol and water bottles on each cage were switched, and the test was carried on for 16 days. The alcohol preference score of a mouse is the ratio of the volume of ethanol solution consumed during the test period to the total volume of fluid consumed. In our hands C57BL/6J mice score 0.79±0.11 (they prefer alcohol), while A/J mice score 0.21±0.09 (they avoid alcohol).

C. Cricket Killing

Cricket killing was tested by introducing a single cricket into the home cage of a mouse, replacing the lid of the cage with a clear plexiglass sheet with small air holes, and counting the time elapsing before the mouse attacked the cricket. C57BL mice will quickly attack and eat the cricket (median latency to attack is 4 minutes for males and 9 minutes for females). A mice, in the same situation, usually fail to attack the cricket at all during the 30 minute time allowed. We cricket tested each mouse twice, on consecutive days, because it has been observed that exposure to one cricket increases subsequent attacking behavior in many mice (Butler, 1973). A mice still fail to attack on the second trial, however. The cricket attacking score of each mouse was calculated as 30 minutes minus latency to attack, divided by 30. Thus the scores range from 0 (for a mouse which failed to attack, i.e., had a latency of 30 minutes) to 1 (for an instantaneous attack, i.e., latency 0).

D. Rope Climbing

The final category of behaviors that we measured was what we call rope climbing. The testing apparatus consists of a clothes line rope stretched vertically between two platforms about two feet apart, so that the mouse can climb up the rope and get off on the top platform, or climb down the rope and get off on the bottom platform. A knot is tied in the rope midway between the platforms. To begin the test a mouse is placed on the knot at time 0. Rope latency is measured as the amount of time elapsed before the mouse leaves the knot and begins to climb. Rope time is the amount of time elapsed after the mouse leaves the knot up to when it steps off onto one of the platforms. The third possibility, rope direction (whether the mouse climbs up or down), was not used in this study because A and C57 mice were found not to differ in this respect. The rope latency is long (1.20±0.7 min) in A mice and short (16±9 sec) in C57. Similarly the rope time is short in C57 mice (1.7±0.2 min) while A mice frequently fail to complete the climb in the five minute trial period.

Altogether we measured seven separate behaviors in these four tests. The open field test gave us *activity* and *boli*. We also collected *alcohol preference, cricket 1, cricket 2, rope latency* and *rope time* scores.

III. THE CHIMERAS

Our series of chimeras comprised 36 individuals, 13 phenotypic females and 23 phenotypic males. All were produced by the technique of aggregation of embryos at approximately the 8-cell stage (Tarkowski, 1961; Mintz, 1962). The chimeras were of two types: A/J<−>C56BL/6J and A/J<−>B10.D2. B10.D2 is C57BL/10J with the H-2 region of DBA/2J. C57BL/6J and C57BL/10J have been found to show the same behaviors in the tests we have used. Sixteen of these chimeras are from a group provided to us by Dr. W. K. Whitten of the Jackson Laboratory. These include all 10 of the A<−>B10.D2 as well as 6 of the A<−>C57BL/6. Fig. 1 shows the distribution of coat color in these chimeras. The A/J strain is albino (*c/c*) while the C57 strains are pigmented (*c+/c+*). Three of the individual chimeras whose coats were albino showed no evidence of C57 cells in any tissue. All of the individuals which were entirely pigmented did have A cells in some tissues.

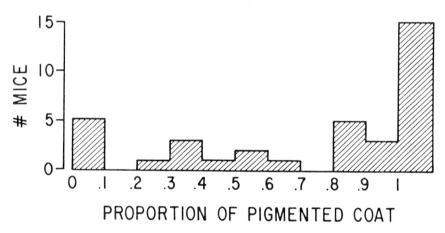

Fig. 1. Distribution of coat color among the chimeras. Values are given in terms of the fraction of the coat showing the C57 (pigmented) phenotype.

IV. BEHAVIORAL VARIABILITY IN CHIMERAS

Since an A<−>C57 chimera is a mixture of A and C57 cells, whether, in a given situation, it behaves like an A mouse, like a C57 mouse, or in some other way must depend upon the way in which the A and C57 cells are distributed in its body. If a given behavior is controlled by a single cell or a clonally related group of cells, then a chimera can only behave like whichever of the parent strains happened to contribute the relevant cell or clone. If, on the other hand, the behavior is controlled by cells derived from two or more clones, then in some chimeras the cell population controlling the behavior will be a mixture of A and C57 cells, and a behavior different from either strain may result.

Fig. 2 shows that behaviors clearly intermediate between those of A and those of C57 mice were evident in our chimeras in all seven of our behavioral measures. This suggests that none of our behavioral phenotypes is controlled by a single cell or clone.

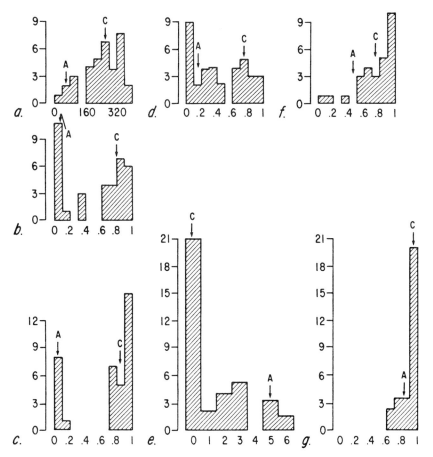

Fig. 2. Distribution of behaviors among chimeras. In each distribution the arrow marked "A" designates the mean score of A/J mice, while the arrow marked "C" designates the mean of C57BL/6J mice.

a) Distribution of number of squares traversed in the open field test.
b) Distribution of scores for latency to attack first cricket. A score of 0 means no attack. A score of 1 would mean instantaneous attack.
c) Distribution of scores of latency to attack second cricket. Scores are as in part b.
d) Distribution of alcohol preference.
e) Distribution of number of fecal boli deposited during the open field test.
f) Distribution of rope climbing time scores. Scores are in terms of the fraction of the total available time (5 min) that was used in climbing.
g) Distribution of rope latency scores. Scores are in terms of the fraction of the total available time (5 min) that elapsed before the mouse began to climb the rope.

V. ASSOCIATIONS AMONG BEHAVIORS

If two or more of our seven behavioral measures were reflections of the same fundamental anatomic or physiologic difference between the A and C57 strains, then those behaviors would be controlled by the same set of cells. In a given chimera that set of cells would have a particular composition (A-C57 mix). All the behaviors controlled by the cell set would reflect this composition, and the scores for these behavior measures should fall in the same region of the A to C57 behavior scale. In other words, there should be a correlation among behavior measures reflecting a given fundamental difference between strains when these measures are made on a series of chimeras.

We applied the technique of factor analysis to our data. This procedure analyzes correlations by clustering variables into factors within which variables are correlated. The computer program BMDP4M was used (Dixon, 1975). The results are factor loadings, i.e., contributions of the individual variables to each factor. Loadings less than 0.250 are not significant here and are replaced by zeros. The method of principal components was used to determine the factors. They were rotated by the varimax method. Table I shows the results of our analysis.

TABLE I

Factor Loading

	Factor 1	Factor 2	Factor 3
Cricket 1	0.975	0	0
Cricket 2	0.972	0	0
Boli	0	−0.836	0.314
Activity	0	0.799	0
Alcohol Preference	0	0.763	0.273
Rope Latency	0	0	0.882
Rope Time	0.282	0	0.794

The data indicate that alcohol preference, open field activity (squares), and boli are highly correlated in our chimeras, and thus may all be expressions of a single underlying strain difference. Latency to attack first cricket is highly correlated with latency to attack second cricket, and rope latency is correlated with rope time. This indicates that our seven measures of behavior may reflect only three underlying mechanisms: one controlling open field activity and alcohol preference, one controlling cricket killing, and a third controlling rope climbing. The loadings of rope time on factor 1 and boli and alcohol preference on factor 3 are only marginally significant. Since we are not certain that these are meaningful or will persist when a larger sample of chimeras is available, we have not included them in our discussion.

VI. DISCUSSION AND SUMMARY

These chimera studies themselves cannot distinguish whether a common mechanism underlying, for example, boli, activity, and alcohol preference operates at the level of the gene, or at the level of the cell. The boli, activity, and alcohol preference behaviors could all reflect a single set of genetic differences between the A and C57 strains, or they could reflect two or three separate sets of genetic differences which happen to be expressed in the same set of cells. There is some genetic evidence which could be interpreted as meaning that the common mechanism underlying these behaviors is at the gene level. First, the albino locus has been shown to exert an influence both on open field activity (Thompson, 1953) and on alcohol preference (Henry and Schlesinger, 1967). Second, Whitney (1972) found a strong correlation between alcohol preference and open field activity in the F2 of a cross between a high activity, high preference strain and a low activity, low preference strain. Whitney reasonably concluded that this could be evidence either for preference and activity being pleiotropic manifestations of the same set of genes, or for linkage between a set of genes controlling alcohol preference and another set controlling activity. Linkage between two distinct sets of genes could not account for the correlation between activity and alcohol preference seen in our chimeras, so the chimera data and the Whitney F2 data taken together tend to favor the interpretation of pleiotropy.

In summary, our work indicates that the differences between the A/J and C57BL/6J strains in alcohol preference, open field activity, defecation in an open field, latency to attack first cricket, latency to attack second cricket, latency to climb rope and rope climbing time are due to three underlying classes of differences. One of these controls cricket attacking (both first and second cricket), one controls rope climbing (both latency and time), and the third controls alcohol preference, open field activity and defecation. These three behavior classes are controlled by three different sets of cells, none of which is clonally derived.

ACKNOWLEDGMENTS

This work was supported by research grants AA 00388 and HD 03015 from NIH to M. Nesbitt, MH 18996 to Karla B. Thomas (now Karla Butler). Computing assistance was obtained from the Health Sciences Computing Facility, UCLA, supported by NIH Special Research Resources Grant, RR-3.

REFERENCES

Butler, K. (1973). *J. Comp. Phys. Psych.* **85,** 243-249.
Dixon, W.J., ed. (1975). BMPD *Biomedical Computer Programs.* University of California Press.
Henry, K.R. and Schlesinger, K. (1967). *J. Comp. Phys. Psych.* **63,** 320-323.
McClearn, G.E. (1959). *J. Comp. Phys. Psych.* **52,** 62-67.
McClearn, G.E. (1960). *J. Comp. Phys. Psych.* **53,** 142-143.
McClearn, G.E. (1972). *Ann. N.Y. Acad. Sci.* **197,** 26-31.
McClearn, G.E. and Rodgers, D.A. (1961). *J. Comp. Phys. Psych.* **54,** 116-119.
Mintz, B. (1962). *Am. J. Zool.* **2,** 432.
Tarkowski, A. (1961). *Nature* **190,** 857-860.
Thompson, W.R. (1953). *Can J. Psych.* **7,** 145-155.
Whitney, G. (1972). *Finn. Found. Alc. Stud.* **20,** 151-161.

III. Plants

Embryo Cells and their Destinies in the Corn Plant[1]

E. H. Coe, Jr.[2]

Science and Education Administration
U.S Department of Agriculture

M. G. Neuffer[2]

Department of Agronomy
University of Missouri
Columbia, Missouri 65211

I. INTRODUCTION

The elaboration of a plant from the zygote through the dormant seed to the adult individual is something we would like to understand on its own merits. In the case of corn, *(Zea mays L.)*, usually called maize for clarity in international usage, understanding the morphogenesis is doubly desirable: Corn is a representative (though highly specialized) monocot in which genetically sophisticated developmental questions may be asked, and the information may provide us paths on which to carry out useful manipulations of this major crop species.

[1]Cooperative investigations of the Agricultural Research Service, U.S. Department of Agriculture, and the Missouri Agricultural Experiment Station; Journal Series No. 7945.

[2]Geneticist, Agricultural Research Service, U.S. Department of Agriculture, and Professor of Agronomy, respectively. Address for both authors: Curtis Hall, University of Missouri, Columbia, MO 65211.

Because plant meristems generate maturing cells in a directional fashion, they are borne at advancing fronts. The apical meristem atop a dicotyledonous stem, for example, is borne forward by the laying down and expansion of differentiating cells maturing behind. The apical meristem of monocots is distinctive: it first lays down a series of intercalary meristems and leaf initials; each intercalary meristem proliferates internodal tissue that elongates to form the stem segments, and each group of leaf initials proliferates an encircling leaf beginning with the tip and continuing until the leaf base is laid down. The result of the monocot pattern of development is a stem in which basal internodes are oldest but each internode is youngest at one end, and each leaf is youngest at its base.

To understand development in the corn plant more fully we would like to know:

Are the organ-forming meristems compartmented?

Are intercalary meristems and leaf initials at the same node derived from identical cells in the apical meristem?

Do clones establish any major parts?

When and how are the reproductive parts set off from the vegetative; are the male and female gametes derived from the same cell lineages?

Are any of these processes pliable or tractable?

Before these questions are considered in the light of some new experiments, a summary of the overall features of this plant's structure and life cycle, along with some of the established details of morphogenesis, will provide a framework.

II. THE CORN PLANT

The fully developed corn plant is a jointed, bamboo-like stem with leaves in an alternate ladder arrangement (Fig. 1). Each leaf, inserted at the node, is an overlapped-cylindrical roll wrapped around the stem, flaring at the top to form the broad, flattened blade portion, which has a thick, stiff midrib in the center. The stem is terminated by the tassel inflorescense, consisting of ten to twenty branches bearing the spikelets, where meiosis occurs and the pollen grains develop. Ear shoots, were meiosis generates the 500 or so embryo sacs, each with an egg nucleus, are branches formed on alternating sides at most nodes up to a midway point on the stem. They have modified leaves (husks) encasing each other and enclosing the compact ear.

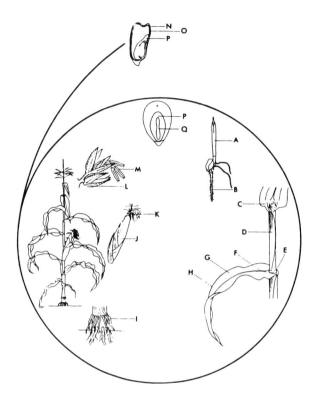

Fig. 1. Morphology and life cycle of the corn plant. Right: leaf, inserted around the node (E), consists of a cylindrical sheath (D) and flattened blade (G) with midrib (F). Left: mature plant at flowering, with the tassel bearing spikelets and anthers (M) and the ear bearing the seeds enclosed by husks (J). (From Styles *et al., 1973*).

III.   THE CORN SEED: FORMATION AND PROLIFERATION

The mature seed consists of both the dormant embryo axis (integrally continuous with the scutellum, a digestive-absorptive cotyledon) and the endosperm containing stored materials (Fig. 2).

The endosperm, a triploid tissue, is derived from a separate fertilization, parallel to that of the zygote, by the fusion of one sperm with two polar nuclei in the embryo sac to form a triploid cell. McClintock shows in this volume the systematic proliferation of this

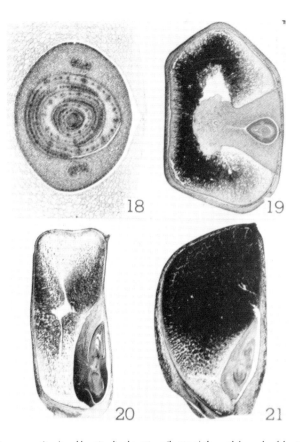

Fig. 2.   Transverse (top) and longitudinal sections (bottom) through kernels of dent (18, 19, 20) and pop corns (21), showing the endosperm and the cotyledon and embryo axis with 5-6 preformed leaves. (From Sass, 1955).

cell, revealed by the mutational events she has studied so intensively. Our picture of this proliferation devolves from McClintock's extensive observations over many years (made almost incidentally in the context of more complex concerns — see McClintock, 1951, 1965) and some confirming data obtained through explicit mutational experiments (Coe, 1977). In the formation of the endosperm, the first division is vertical and sets off left and right halves; the second division is in a vertical (somewhat diagonal) plane perpendicular to the first. After another 1-3 divisions, proliferation proceeds spherically by cambium-like expansion in cones and fans (McClintock, 1951 and this volume) and ends with systematic divisions in alternating planes at the surface (Coe, 1977; Fig. 3).

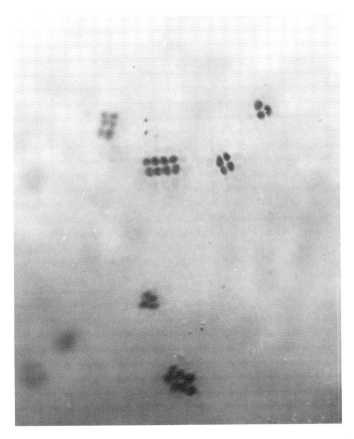

Fig. 3.   Alternating-plane clones of cells proliferated in the last few divisions in the surface layer (aleurone tissue) of the endosperm, revealed by pigment formation following mutation. (From Coe, 1977).

The embryo in the dormant seed has 5-6 leaves already formed waiting to expand (Fig. 2), and a meristematic area, the shoot apex, from which about 15 more leaves and the ears and the tassel will be elaborated. By the time the seedling has grown for 4-5 weeks all of the parts have been set out, including the tassel and ears (Fig. 4), in a stepped pyramid of intercalary meristems, nodes and internodes.

Fig. 4   Section (left) and dissection (right) of plant stem and apex at 32 and 35 days after planting, respectively, with stepped pyramid of intercalary meristems, rudimentary tassel and rudimentary ear branches. (From Kiesselbach, 1949).

Because the more advanced leaves surround the less advanced, each young leaf must come out from inside the previous one, telescoping, resulting in the famous summer-night "growth popping" sound due to frictional release. The half-foot high structure expands and telescopes fifteen-to-twenty fold during the subsequent 5-6 weeks.

The details of early morphogenesis of the embryo have been described by Randolph (1936; Fig. 5). The zygote divides twice in horizontal planes, laying down the suspensor beneath and the embryo initial above (Fig. 5, C-F). The first vertical division of this initial (Fig. 5, G) at about 36 hours after pollination has been shown by Steffensen (1968) to separate left and right halves of the plant, dividing each leaf in the center at the midrib and continuing all the way up the midline of the tassel. Steffensen identified this compartmentation by marking genetically for losses induced by radiation during the first few divisions. In this instance the compartments and the clones share the same boundary. Fig. 6 shows a young plant in which this division was differential for albinism after random loss of a ring derived by McClintock to cover the *wd* (white) deficiency (McClintock 1944). As Steffensen (1968) has put it, "Many of the aspects of maize development presented in this manuscript were already known to Dr. Barbara McClintock more than (30) years ago."

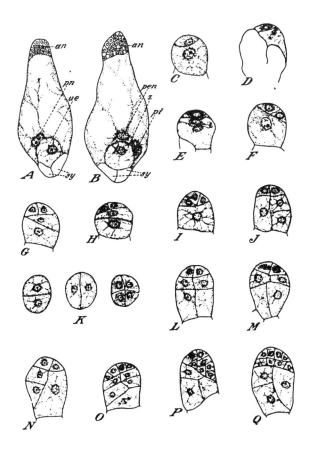

Fig. 5.   Early morphogenesis of the corn embryo. A, unfertilized embryo sac; B, fertilized embryo sac; C, D, 2-celled stage at 32 hours after pollination; E-G, 3 and 4-celled stages at 36 hours; H-J and L-N, longitudinal sections at 42 hours; K, cross sections from base to tip of a stage similar to H; O-Q, longitudinal sections at 4 days. The embryo develops from the smaller, more dense cells at the top. (From Randolph, 1936).

Fig. 6.   Sectored plant in which loss of an unstable ring chromosome leading to failure of green pigmentation chanced to occur at the early division separating left and right halves of the plant.

From the dormant embryo forward, Stein and Steffensen (1959a, b) have defined the morphogenesis and growth of the first 5-6 leaves; Steffensen has developed a detailed reconstruction of the shoot apex and of the generation of the 15 or so additional leaves from it. Each cell of the dormant shoot apex (Fig. 7) elaborates into clones (i.e., files of cells); following irradiation-induced genetic losses, sectors are elaborated that reveal radial proliferation into a sequence of clonally related leaf segments (Fig. 8). Sectors have, as a result, a systematic pattern of location in successive leaves (Fig. 9) that depends on the location of the cell in the perimeter of the apex. We only have a little new information to add to Steffensen's neat picture: The shoot apex is further subdivided.

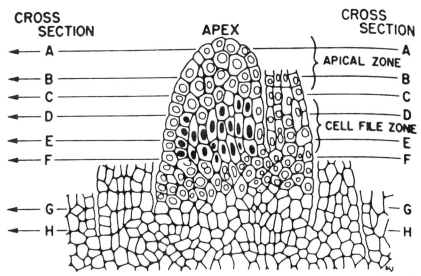

Fig. 7. Longitudinal section through the shoot apex of the dormant embryo, including leaf 5 (appressed to the right flank and encircling the dome), portions of the four other leaves, and the cell geometry in the apex; "cross section" refers to precise photographs and drawings of sections given in the original paper. (From Steffensen, 1968).

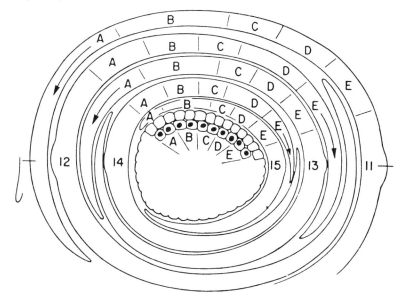

Fig. 8. A reconstruction in cross section of the pattern of growth of cells from the shoot apex (refer to Fig. 7), leading to leaves 11-15 by radial proliferation of single cells (A, B, etc.) into related leaf segments. (From Steffensen, 1968).

Fig. 9. Sectors in a plant of *cm* (chloroplast mutator) genotype, in which chance mutations in plastids (Stroup, 1970) develop clonally and show the characteristic patterns of location defined by radial proliferation of cells in the shoot apex (see Figs. 7 and 8).

Fig. 10. *A B Pl* (left) and *A b Pl* (right) plants at the ear level, showing the difference in expression in these tissues for one of the four factor pairs used to trace cell lineages.

IV.   CELL DESTINIES

The experiments to be described were conceived and conducted because we recognized that we did not know a simple fact — whether the ears and tassel had any cells in common in the embryo. Mutational treatments applied in the embryo stage either could or could not (with very different genetic consequences) lead to a mutant clone destined to be represented both in male gametes and in eggs, but which was true?

To examine the proliferation of the embryo cells into the plant body and the reproductive organs, we used genetic marking with anthocyanin factors that would permit somewhat more extensive tracing of sectors than the chlorophyll markers employed in previous studies. In the

Fig. 11.   Sector proliferated after a genetic loss induced in an embryo cell by irradiation, extending over four nodes in the sheath and stem (culm).

Fig. 12.   Sectored ear branch.

Fig. 13.   Right, two sectored husks in sequence, from the ear branch shown in Fig. 12.
Fig. 14.   Left, the ear from the branch shown in Fig. 12: the light brown portion is from the mutant clone, the dark purple from non-mutant.

presence of the dominant genes A, B, Pl and $R\text{-}r$, the mature plant is purple in virtually all parts (see Fig. 10). Dry seeds heterozygous for these four markers, A a B b Pl pl $R\text{-}r$, $R\text{-}g$, were x-rayed (10,0000 r) and grown. Depending on the tissue (Coe and Neuffer, 1977), two or three of the dominant alleles are required for color, and mutation or loss of one of them leads to a green sector on a purple background. Such green sectors, the majority of which would be losses of the chromosome arm carrying the dominant gene, were scored in approximately 500 treated plants for their extent (number of nodes affected) and width as a percentage of the perimeter (Figs. 11, 12). The ear-branch sector shown in Fig. 12 continued from the husks (Fig. 13) into the ear proper (Fig. 14).

The numerical data for 45 tassel sectors (Fig. 15) show that the fraction affected (affected branches divided by total branches) was near 50% for most of the sectors — i.e., the apparent number of cells (the reciprocal of the fraction) was two for most tassels. Many sectors, however, affected less than one half, with apparent cell number (ACN)

TASSEL: FRACTION AFFECTED

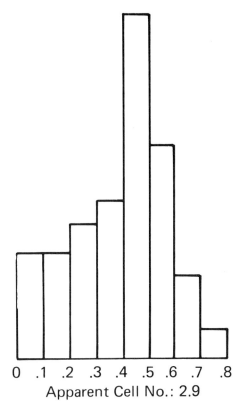

Apparent Cell No.: 2.9

Fig. 15.   Frequency distribution of 45 tassel sectors according to the fraction affected.

of 3, 4 or 5 cells; the average ACN was 2.9 cells. Anderson *et al.* (1949) calculated (for different material) that 7 or 8 cells represent the tassel in the dormant seed, from sectors showing pollen sterility following atomic irradiation. Over half of the tassel sectors in the present experiment divided the central spike vertically and the tassel branches into two symmetrical halves, consistent with two cells whose separated lineages trace to the first vertical division of the zygote defined by Steffensen (1968). In other words,these two to four cells are set off early as a sample of the whole plant-to-be; there is no "germ line" here in the usual sense.

The numerical data for 94 sectors in the body of the plant can be summarized according to the average ACN at each node level and the average extent in number of nodes (Fig. 16). Approximately 16 cells constitute the upper nodes, then 32 cells below, with the average extent changing from 5 or so nodes above the ear level to 2 or 3 in the lower portion. Since the extent of each sector varies widely, each level usually including sectors extending for 1, 2, 3 or 4 nodes (often more), each cell in the embryo may be free to proliferate to a variable extent rather than to a fixed extent; assuming this to be correct the cells are quite independent components of a level rather than restricted by the specifications of their level.

A simplistic interpretation of the sectoring data (Fig. 17) identifies the classes just derived: 2-4 cells destined to become the tassel; 16 cells to become the upper nodes, extending 4-7 nodes; 32-cell classes three deep, each class extending 2-3 nodes; two subsets (one on each side of

BODY SECTORS

Node Level	Apparent Cell No.	Extent (Nodes)	
19+	21.0	5.5	
18	15.3	8.5	
17	16.4	5.5	
16	17.1	5.0	
15	21.8	3.4	
14	33.4	2.8	Ears
13	30.6	2.3	
12	30.0	2.0	
11	31.7	2.3	
10	31.9	3.0	
9	36.3	2.0	
8	40.3	2.0	
7	19.0	2.0	

Fig. 16.   At each node level, the apparent cell number (reciprocal of fraction affected), extent in nodes, and location of ears for 94 sectors in the plant body.

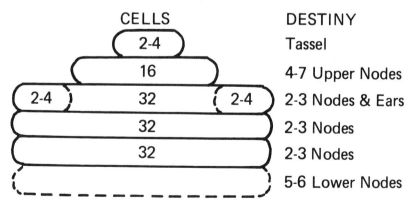

CELLS	DESTINY
2-4	Tassel
16	4-7 Upper Nodes
2-4 32 2-4	2-3 Nodes & Ears
32	2-3 Nodes
32	2-3 Nodes
	5-6 Lower Nodes

Fig. 17.   Simple interpretation of classes of cells in the shoot apex of the dormant corn embryo.

the 32-cell class) of 2-4 cells each destined to enter the ear branches *from a body sector*. The diagrammatic interpretation agrees reasonably well with anatomical descriptions of the shoot apex in the dormant embryo (Fig. 7), including a rough approximation to the estimated cell numbers in the meristem (Steffensen, 1968). A few events observed in tillers (basal branches) affected one-half to one-fourth of the branch (extending from the basal nodes through the tassel), indicating that two to four cells form the entire branch just as the two cells in the first vertical division of the zygote form the main stalk; at least some tiller sectors were expressed also in the lower sheaths of the main stalk.

The setting-off of intercalary meristems to form the nodes and internodes in monocots, strategically effective in dividing responsibilities toward rapid extension of the shoot apex, is evidently preceded, at least in this determinate monocot, by dividing into classes that will become two to several intercalary meristems. This is a further strategic division of responsibilities.

Since each of the classes in Fig. 17 is divided, in the plane of the page, into left and right halves, it is possible that each ear, like the tassel, is normally derived from one or two cells from each half. In the only analyzable ear sector found in these experiments (Fig. 14), the cell in which the mutation occurred gave rise to the majority of the ear (over 95%), suggesting that there is flexibility (and thus a potential for somatic selection) in the elaboration of the ear from its initials. The elaboration of the tassel and body parts, however, appears to be much less flexible (at least in the left-right sense), since well-defined compartments for them are evident.

Compartmentations in the corn plant can be seen to include the following levels: (1) an early separation for left vs. right in which the compartment is the clone; (2) a separation for tassel vs. upper section vs. lower sections in which two to four cells (clones) are destined to the tassel compartment only while larger numbers of clones are destined to body sections; and (3) paired separations for ear branches (including part of the plant body) vs. the plant body (only). While clonal proliferation has a role in specifying compartments in the first separation (left-right), it overruns the second and third separations. Thus, neither compartmentation nor clonality dominates consistently as the determinative process in the corn plant; clonality, however, obviously plays a major role in influencing the elaboration.

V. SUMMARY

The questions posed in the introduction can be given summarized response:

The meristems are compartmented, both early in embryogenesis and late in embryogenesis, to form groups of organs or parts of organs.

Intercalary meristems and leaf initials derive from parent cells they have in common in the apical meristem of the dormant embryo.

Clones in some instances establish major compartments such as the left-right separation (and half-tassels), but the intercalary meristems partition clones, apparently without regard to their extent.

The reproductive parts in the tassel are set off from the vegetative body in the dormant embryo, but the ear branch is from a subset of the cells that form the vegetative body sections; the male and female gametes are set off from each other by the sectional compartmentation (Fig. 17) that occurs prior to dormancy.

Pliability and tractability of these processes remain to be explored; the methods and background are at hand for study, and cursory observations indicate that plasticity (e.g., accommodation for killed cells) can be studied with facility in this material.

Further exploration of the events following germination, including especially the specification and formation of the intercalary meristems and leaf initials, should derive a more adequate picture of these processes than is presently available. It will also be of interest to relate cell lineages to the clonal distributions of organelle-inherited ("cytoplasmic") properties; initial observations suggest that at least some unstable cytoplasmic traits (chloroplast mutator and non-chromosomal stripe) are fully consistent with the cell lineages while others (e.g., iojap) include both cell-lineage transmission and patterned, bilaterally symmetrical distributions.

ACKNOWLEDGMENTS

We appreciate the expert and alert assistance of Sheila McCormick in the experiments, and the enthusiastic interest and critique of Virginia Walbot during this study.

REFERENCES

Anderson, E. G., Longley, A. E., Li, C. H. and Retherford, K. L. (1949). *Genetics* **34,** 639-646.

Coe, E. H., Jr. (1977). *Proc. Symp. Maize Genet. and Breed.,* Wiley Press, New York (in press).

Coe, E. H., Jr. and Neuffer, M. G. (1977). The genetics of corn. *In:* "Corn and Corn Improvement" (G. F. Sprague, ed.), 2nd ed., Am. Soc. Agron., Madison, Wisconsin (in press).

Kiesselbach, T. A. (1949). *Nebraska Agric. Exper. Sta. Res. Bull.* **161.**

McClintock, B. (1944). *Genetics* **29,** 478-502.

McClintock, B. (1951). *Cold Spring Harbor Symp. Quant. Biol.* **16,** 13-47.

McClintock, B. (1965). *Brookhaven Symp. Biol.* **18,** 162-184.

Randolph, L. F. (1936). *J. Agr. Res.* **53,** 881-916.

Sass, J. E. (1955). Vegetative morphology. *In:* "Corn and Corn Improvement" (G. F. Sprague, ed.), pp. 63-87. Academic Press, New York.

Steffensen, D. M. (1968). *Am. J. Bot.* **55,** 354-369.

Stein, O. L. and Steffensen, D. M. (1959a). *Z. Vererbungsl.* **90,** 483-502.

Stein, O. L. and Steffensen, D. (1959b). *Am J. Bot.* **46,** 485-489.

Stroup, D. (1970). *J. Hered.* **61,** 139-141.

Styles, E. D., Ceska, O. and Seah, K. T. (1973). *Canad. J. Genet. Cytol.* **15,** 59-72.

Ontogeny of the Primary Body in Chimeral Forms of Higher Plants

Robert N. Stewart

Research Horticulturist
Florist and Nursery Corps Laboratory
Beltsville Agricultural Research Center
Science and Education Administration, Federal Research
U.S. Department of Agriculture
Beltsville, MD 20705

I. INTRODUCTION

The primary body of higher plants consists of their apical meristems and the root, leaves, nodes, branches, and flowers more or less directly derived from them. Most monocotyledons and herbaceous dicotyledons complete their life cycle with this primary growth. Secondary growth is produced by activity of a secondary vascular cambium which increases the amount of vascular tissues and results in a thickening of the shoot axis. The term 'shoot' is used to include all the primary body except the root.

Shoots grow in length and add new tissues and organs through formation of new cells in a layered apical meristem. Cells in the distal region of the apex are arranged in layers parallel to the surface (Fig. 1). The one or more peripheral cell layers are maintained by almost exclusive anticlinal divisions and have been called the tunica. The core of cells whose divisions are in random planes is called the corpus. The independent, clonal nature of the cell layers and their derivatives in ontogenesis was established by studies of periclinal cytochimeras by Satina et al. (1940), Dermen (1945), and others. Satina et al. (1940) initiated the convention of designating the outermost apical layer as L-I and the next inner layer as L-II. The innermost clone of cells, the corpus, was also called a layer, L-III. Baur (1909) suggested that the repeated pattern of green and white tissue in the leaves of a geranium periclinal chimera must reflect differences inherent in layers of cells in the shoot apex. There have apparently been no observations of differences between the structure of mutant and wild type proplastids in the apical layers. However, Dermen (1947) and Stewart and Dermen (1970a) have pointed out the very similar pattern of mutant tissue in cytochimeras and plastid chimeras. It therefore seems reasonable to assume that there are apical layers in plastid chimeras which differ in their genetic potential for plastid development. It should be noted that epidermal cells, except for the guard cells of stomata, do not develop normal chloroplasts. Therefore, a genetically green (G) L-I does not contribute

Fig. 1A — Longitudinal section of shoot tip of colchicine treated American elm. Only the second apical layer, L-II, has become tetraploid (4N). The outer layer, L-I, and the inner layer or corpus, L-III, have remained diploid. This apical meristem would be described as 2N-4N-2N. The portion shown distal to the most recently formed leaf initial is called the apical dome. Fig. 1B — 2N-4N-4N elm cytochimera. Fig. 2 — Chimeral cotton seedling from the cross of a virescent haploid-producing line as seed parent and a green tester line as pollen parent. The pale sector is haploid virescent and the dark sector haploid green.

to the color of stem or leaf.

We have suggested (Dermen, 1960; Stewart and Dermen, 1970a; Stewart et al., 1974) that gymnosperms basically have two apical layers and angiosperms three. Histological sections of apices occasionally show 4 or 5 layers of cells and there are reports of derivatives of up to 5 layers in stem and leaf (Dermen, 1951, 1953; Stewart et al., 1974; Stewart and Dermen, 1975). However, the inner layers were usually replaced and left behind within growth of a few nodes. A number of workers have observed more frequent periclinal divisions of cells within internal layers (Dermen, 1951, 1953; Clowes, 1961; Thompson and Olmo, 1963; Tilney-Bassett, 1963; Pratt et al., 1967) resulting in the relative impermanence of inner layers.

In plant literature the term "chimera" is used to describe a shoot with genetically different clones, whether originally from a single zygote or from a mixture of cells from two zygotes. The term "mosaic" was first used to describe a variegated pattern of mixed small areas of green and white leaf tissue in a plastid chimera. It has since been used to describe similar variegation resulting from virus infection. The only persistent plant chimeras are periclinal chimeras. A periclinal chimera with a sector of a different cell lineage within a layer is called a mericlinal chimera.

Many botanists have been reluctant to utilize chimeras in onto-genetic studies (Boke, 1948; Clowes, 1956, 1961; Gifford and Corson, 1971; Domergues, 1962; Soma and Ball, 1964; Nougarede, 1967). The growth and differentiation of many chimeral plants is clearly distorted and doubts were expressed as to whether any chimeras could be used to interpret normal ontogenesis. However, in a number of chimeras (Stewart, 1965; Stewart and Arisumi, 1966; Stewart and Dermen, 1970a; Stewart et al., 1974) the amount of tissue in an organ from a given mutant clone was essentially the same as from a normal clone regardless of its position in the meristematic regions. We believe that the validity of this type of material has been fully established and that chimeras are an indispensable tool for research in experimental developmental biology.

II. SOURCES OF CHIMERAL PLANTS

A. *Somatic Mutation*

The origin of most chimeral plants has been most certainly through spontaneous somatic mutation. The maintenance of many chimeral forms over many years is possible because of the indeterminate growth of their vegetative apical meristem. Many of the most important florist and nursery crops are asexually propagated and most of the important cultivars are chimeras which originated as somatic mutants. At least 90 percent of the standard carnations now grown are chimeral forms of the original red 'Wm. Sim' which was of seedling origin in 1935 (Farestveit, 1969). Periclinal chimeras are common in chrysanthemums, roses, poinsettias, and many other asexually propagated crops.

B. *Grafting*

A second source of chimeral forms has been from grafting, i.e., joining a scion from one seedling to a root stock from another. This is a common practice in the propagation of many ornamental and crop plants, especially roses, apples, and other woody species. While a grafted plant might be considered a chimera, the only association between the components is in a limited area at the graft union and no information relative to ontogenetic development is provided. A Florence garden in 1644 was the site of the first recorded true chimera from grafting. Apparently a graft between sour orange and citron failed, but a bud grew from callus in the area of the unsuccessful union. The resulting shoot was different in appearance from either scion or stock and has been propagated asexually ever since as the 'Bizzarria Orange'. Some branches and fruits show sectors which are typically orange or citron and 'Bizzaria Orange' is now considered to be a periclinal chimera with a layer of orange cells over a core of citron tissue. Winkler (1907) grafted tomato and nightshade in an attempt to produce a graft hybrid. After cutting off successful grafts through the area of the union, he obtained adventitious shoots which were intermediate between scion and stock. Occasional branches showed sectors of both parental types and Winkler used the term chimera to describe them. In the next 20 years a number of other workers (Neilson-Jones, 1969) produced and described a number of graft chimeras of several Solanaceous species. Recently, Stewart et al. (1972) described many differences between the cell lineages in a periclinal graft chimera of *Camellia japonica* and *C. sasanqua*.

C. *Changes in Chromosome Number*

Another source of chimeras has been the doubling of chromosome number with colchicine or other chemical or shock treatment. Cells at different stages of the division cycle are differentially sensitive so that in a treated apical meristem only a limited number of cells are affected.

If certain cells are affected and continue to divide, a cytochimeral form (Fig. 1) may be established and persist (Burk, 1975).

D. *Mixed Callus Cultures*

Another promising source of chimeral plants is the formation of adventitious chimeral shoots from mixed callus in tissue culture. Carlson and Chaleff (1975) have described a number of arrangements of tissue in adventitious shoots from mixed calli of two tobacco species. Ramulu et al. (1976) found cytochimeras in *Lycopersicon peruvianum* plants regenerated from *in vitro* cultures of anthers and stem internodes.

E. *Somatic Crossing Over*

Chimeral sectors and branches have been found in cotton plants, apparently resulting from somatic crossing over or chromosome deletions (Barrow et al., 1973; Barrow and Dunford, 1974). The "homozygous" lines are allotetraploid and the twin spots or sectors are a darker or lighter green. Somatic crossing over in plants establishing different cell lines as twin spots or sectors has also been reported by Jones (1937), Vig (1973) and others.

F. *Semigamy*

Another source of chimeral plants is the haploid-producing semi-gametic cotton lines of Turcotte and Feaster (1967). Pollination of the haploid-producing line with various tester lines produces many chimeral progeny with separate haploid clonal cell lines of the parental genotypes and/or the expected fusion genotype (Fig. 2).

G. *Plastid Segregation*

Many green and white variegated chimeral plants have been shown to be the result of mutation in the chloroplast DNA. Cells with mixed plastid types have been reported by many workers (Gregory, 1915; Woods and DuBuy, 1951; Michaelis, 1958; Burk et al., 1964; Stewart, 1965; Tilney-Bassett, 1963). Maternal inheritance is usually observed although several cases of male transmission have been reported (Tilney-Bassett, 1963; Stewart et al., 1974). An important source of new chimeral arrangements are the heteroplastidic zygotes which result from male transmission in crosses between homoplastidic mutant and normal parents or from the formation of heteroplastidic female gametes in plants with heteroplastidic clones giving rise to the gametes (Burk et al., 1964; Stewart, 1965). The segregation to homoplastidic clones in the various apical layers of heteroplastidic seedlings eventually gives rise to all possible chimeral arrangements of mutant and normal tissue in the apical meristems.

III. RECOGNITION OF CLONAL DIFFERENCES

A. *Chromosome Number and Cell or Nuclear Size in Cytochimeras*

Clonal cell lines of different chromosome number were found in the graft chimeras of tomato and nightshade and this evidence was final proof of their chimeral rather than hybrid nature (Lange, 1927). Jorgenson and Crane (1927) described cytochimeras arising from wound callus. Satina (1945), Satina and Blakeslee (1941, 1943), Satina et al. (1940), and Dermen (1945) provided early reports of cytochimeras resulting from colchicine treatment and their usefulness in developmental studies although they made few observations of mature differentiated tissue. Clonal lines of different ploidy are most easily distinguished in young meristematic tissue, especially in the apical meristem where there are frequent mitotic figures and uniform cell and nuclear size (Fig. 1). In differentiated regions care must be exercised in determining ploidy differences based on cell and nuclear size, but some useful information can be obtained.

B. *Cellular Phenotypes Controlled by Nuclear Genes*

Many clonal cell lines in plant chimeras differ by mutation in chromosomal DNA, although this may be impossible to determine if the mutant clonal line is not involved in gamete formation. An example is a somatic mutation in the epidermis of 'Red Wing' azalea flowers. The mutation suppresses the synthesis of quercetin and quercetin 5-methyl ether glycosides which are copigments with cyanidin 3,5-diglucoside in 'Red Wing' azalea. The mutant petal epidermis is a more orange-red color (λ_{max} 507) than 'Red Wing' (λ_{max} 518) (Asen et al., 1971). The absorption spectra of a mutant epidermal cell in the chimera and of an adjacent wild-type hypodermal cell show a loss of absorption in the 350 nm region due to loss of the quercetin copigments as well as the shift in the visible λ_{max} (Fig. 3). A color mutation from red to pink in poinsettia (Stewart and Arisumi, 1966) was found to be due to suppression of pigment in the epidermal layer (Fig. 4). This mutant later moved into

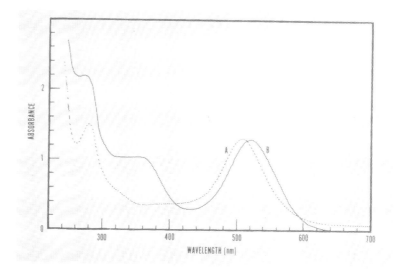

Fig. 3 — Absorption spectra of single living cells found adjacent to each other in the petals of a chimeral azalea. Cells in the mutant epidermal layer (A) show a shift in the visible absorption to shorter wavelengths and the loss of absorbers in the 350 nm region. The spectrum of an adjacent subepidermal cell (B) is typical of those of cells in the original 'Red Wing' variety.

L-II where it gave rise to germ cells and was shown to be a Mendellian recessive. The homogeneous homozygous recessive is white and apparently a mutation at the same locus as the white allele previously known. These cell lines can be identified only in differentiated tissues which normally synthesize color. In chrysanthemum (Stewart and Dermen, 1970b) sectors and branches chimeral for cell lines with yellow chromoplasts are formed when a chromosome carrying a suppressor gene is lost in aberrant mitoses. Many other pigment chimeras have been described without clear evidence of the type of genetic change in the mutant clonal cells.

Fig. 4 — Bracts from a red (R) poinsettia seedling and from the chimeral forms derived from a single somatic mutation suppressing pigment synthesis. Red bract is R-R-R, pink bract W-R-R, pink bract with white edge W-W-R, and white bract W-W-W. Gametes are derived from L-II so that on selfing R-R-R and W-R-R produce red seedlings and W-W-R and W-W-W produce white seedlings. Roots on cuttings are from L-III, thus all but W-W-W produce red adventitious shoots from roots. W-W-R produces white seedlings and red shoots from its roots. Fig. 5 — Sector of red R-R-R composition on a W-R-R poinsettia chimera. The sector occupies one-third the circumference of the stem.

C. Cellular Phenotypes Controlled by Plastogenes

Plastid chimeras are especially useful in developmental studies because they can readily be seen both macroscopically (Fig. 6) and microscopically (Fig. 7). In mature tissue of herbaceous periclinal plastid chimeras the complete cell lineages are clearly recorded. There is a persistent, precise, visible record of the orientation and sequence of cell divisions during development of the primary body. Macroscopic observations of change in pattern makes it possible to limit microscopic observations to areas where significant events have occurred (Fig. 7). It should be noted that not all mutant clones are useful for developmental studies. Mutant clones may be more or less vigorous than wild types and comparison of their relative contribution to the primary body from various chimeral arrangements should be considered (Stewart et al., 1974).

D. *Lineages with Multiple Differences*

The graft chimeras are of particular interest because they provide clonal lines which differ in many genetic characteristics. The only such chimera now available in the U.S. was described by Stewart et al. (1972). Many characters expressed and measured at the cellular level showed complete independence and made it possible to follow cell lineages precisely. Some characters expressed at the organ or whole plant level showed an interaction between the two components.

 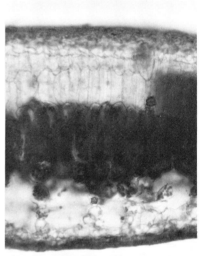

Fig. 6 — Leaf from a G-W-G plastid chimeral tobacco shoot. The dark tissue is derived from the green L-III. The changes in shade are from shifts in the number of cell layers derived from L-II or L-III following intercalary periclinal cell divisions.

Fig. 7 — Cross section of a English Ivy leaf through the shift from a dark green to a lighter green area like those shown in Fig. 6.

IV. STABILITY OF CHIMERAS

Periclinal chimeras are excellent material for ontogenetic studies for a number of reasons. First, they persist indefinitely and provide any amount of material for observation and dissection. There are many different chimeras available whose growth is quite normal with apparent complete compatibility of clearly marked independent clonal cell lineages. The stable homoplastidic chimeras provide a precise picture of the position derivatives of the apical layers take in the primary body. Secondly, few if any chimeras are completely stable. Periclinal divisions in the apical layers or any place in the differentiating primary body produce sectors which mark the event and reveal the subsequent cell lineage. When a cell in L-I or L-II divides periclinally the internal daughter cell may form a layer of cells, replacing and leaving behind L-III. If L-II or L-III cells divide periclinally the outer daughter cell may intrude between cells of the next outer layer, displacing them and thus leaving them behind. Chimeras differ in their stability (Stewart, 1965; Stewart et al., 1974) but in many the frequency of sectoring is high enough to provide adequate data (Stewart and Dermen, 1970a; Stewart and Burk, 1970), but not so frequent as to confuse the consequences of separate events. In some material the large number of sectors produced by segregation in heteroplastidic clones (Fig. 8), mutable genes or mutagenic agents make interpretation difficult because of overlap or blending of the marked sectors. Displacement and replacement are events which occur in normal development as a result of periclinal divisions. When they occur in a chimera, a cell lineage is revealed with no disturbance of normal growth.

Fig. 8 — Poinsettia chimera in which L-II and its derivatives are heteroplastidic for a mutant yellow plastid. The many small sectors are the result of plastid segregation in mitosis (Burk et al., 1964). The events establishing new clones are so frequent that cell lineages are fragmented and developmental patterns are obscured.

V. ONTOGENY OF THE PRIMARY BODY

A. *Apical Initial Cells, the Ultimate Source of New Growth*

Esau (1965) has defined an initial cell as a cell that divides into two daughter cells, one of which remains in meristem while the other is added to the meristematic tissues which eventually form the primary body. The domed shape of the apex suggests that for a cell to remain an initial cell it must remain distal to all other dividing cells within its layer. Otherwise, the divisions of intervening cells would soon leave it behind in differentiating tissue. The unique property of initials is that they continue indefinitely to contribute new growth. Therefore, changes in a shoot which persist in new growth can be presumed to be the result of changes in the initial or initials. A change in a cell in the proximal region of the meristem would appear only in that cell and the few daughter cells from its final divisions before differentiation. A cell on the shoulder of the apical dome would produce more daughter cells and sector originating there would be wider and longer, but would still be left behind as more distal cells continued to divide. Changes in cells closer to the initial cells would result in longer and wider sectors but would still be left behind as the initial continued to divide. If there were a single initial cell in each layer as the ultimate source of new cells, a change in it would be seen in all new growth. If there were 2 or more initial cells, changes in them would result in persistent sectors encompassing 1/2, 1/3, 1/4 or less of the circumference of the shoot.

The event in our chimeral plants which produces a marked sector is the periclinal division resulting in replacement or displacement of the adjacent apical layer. The changes we have observed which persisted along many nodes in the stem were complete changes or broad sectors covering 1/2 or 1/3 (Fig. 9, 10) of the stem (Stewart and Dermen, 1970a). We interpret this as conclusive evidence that there are from 1 to 3 initial cells in each apical layer which are actively dividing and the ultimate source of new growth.

Many workers have stated that there are no specialized or different initial cells in the apex, and that cells function as initials only because of their position on the dome (Soma and Ball, 1964; Newman, 1965; Esau, 1977). They further suggest that, because of changes in shape of the dome during formation of successive leaf primordia the distal position is occupied by different cells, i.e., there is a constant shifting and changing of the cells which are initials (Newman, 1965; Wardlow, 1968). We suggest that there are relatively permanent resident initial cells at

Fig. 9 — Mericlinal plastid chimera in poinsettia. The chimeral sector is G-W-G and occupies one-half the circumference of the stem. Fig. 10 — Mericlinal plastid chimera in juniper which has only two apical layers. The dark sector is W-G and the white sector occupying one-third of the shoot is W-W.

the distal point. This is evidenced by persistent sectors through many nodes. Bain and Dermen (1944) and Dermen (1945) followed mericlinal cytochimeral sectors along 1.3 m of cranberry stem through approximately 100 nodes. Stewart and Dermen (1970a) followed sectors in a plastid chimera through 104 nodes in juniper (Fig. 10), 26 nodes in privet, and 50 nodes in poinsettia. Any one of the initial cells in a layer may lose their position at the distal point and be left behind, as evidenced by the shift in width of persistent sectors between 1/3, 1/2 and complete change (table 2 of Stewart and Dermen, 1970a).

Persistent sectors in periclinal chimeras can be used to determine rate of division of initial cells while other techniques cannot distinguish between the mitotic index of the initials and of their most recent derivatives. In privet (Stewart and Dermen, 1970a) sectors which were less than 1/3 of the circumference of the stem averaged from 3 to 5 nodes in length. Thus we conclude that one division of an initial cell is needed to form approximately 4 nodes of growth through further divisions of the daughter cell which did not remain an initial. Plastochron length in privet is approximately 3 days during vigorous vegetative growth and therefore the initials divide once in 12 days (288 hr). Assuming a mitotic span of 4 hr the mitotic index would be 1.4 percent. This rate of division is adequate to provide the ultimate source of new cells in vigorous vegetative growth of privet. To observe the mitosis of one initial cell would require searching 72 median longitudinal sections of apical meristems. The many reports of no divisions or very few divisions at the shoot apices (Buvat, 1955; Savelkoul, 1957; Popham, 1958; D'Amato and Avanzi, 1968; Lyndon, 1976) are not convincing evidence of a quiescent zone which makes no significant contribution to the vegetative growth of the shoot. The persistent mericlinal chimeras dictate divisions of initials as their ultimate source.

To observe periclinal mitoses of initial cells would be an even more staggering task. During growth of about 20,000 nodes in a population of chimeral California privet, there were about 5,000 normal anticlinal divisions of the initials. As we have shown, one division of an initial results in 4 nodes of growth produced by the daughter cell which does not remain an initial. In this chimeral privet material, 22 shoots developed wide, persistent sectors, the result of 22 periclinal divisions of initials. However, to observe 22 periclinal mitoses of initial cells cytologically would require looking at median longitudinal sections of approximately 350,000 apices.

Esau et al. (1954) stated, "The deciding criterion in definition of initials seems to be not the abundance of mitoses and the amount of contribution to the plant, but the function of cells as the ultimate source of cells and tissues composing the plant. The proper question, therefore, is: how many cell divisions would suffice to make apical cells the source of all other cells of the shoot in primary state . . .? The most crucial evidence of the existence of apical initials would be supplied if such presumptive initials were tagged in some way and the new character then propagated through the part of the plant body derived from these initials." In our material, initial cells are tagged as a result of their rare periclinal divisions. The new sector then marks clearly and precisely the part of the plant body derived from the initial.

B. *Axillary Buds*

Lateral buds are initiated later than leaves, usually in the axis of the second or third leaf primordium below the apex. The buds soon develop a visibly layered cell arrangement in their apical meristem. Histological studies of the origin and development left some doubt as to the relationship of the bud apical layers to those of the shoot on which they arose. But chimeral shoots have chimeral branches (Fig. 11). The fact that branch apices on periclinal chimeras maintain the hierarchy of apical layers of the terminal apical meristem means that derivatives of the apical layers have maintained their position down through the region of leaf initiation. It also means that the cell divisions in the outer layers of the stem during initiation of the lateral bud are anticlinal relative to the new apex. Most periclinal chimeras are stable in maintaining their apical layers in lateral buds even under conditions which alter the contributions of the apical layers to leaf development (Stewart and Dermen, 1975).

Fig. 11 — G-W-G plastid chimera of chinese privet. The shoot on the left was grown continuously at 26 C. The shoot at the right was grown at 16 C and shifted to 26 C. More tissue was derived from L-III at 16 C than at 26 C. However, the hierarchy of layers in the terminal apical meristems or in the lateral buds was not affected, as indicated by the appearance of the leaves formed after the shift in temperature and the appearance of the leaves on the lateral branches forced to growth by later removing the shoot tip. Fig. 12 — Plastid chimeral shoots of geranium. The upper shoot is G-G-W and has a white stem and petioles. The lower shoot is G-W-G and has a green stem and petioles.

C. Stems

Most higher plants maintain a 3-layered apical meristem. Most of the primary tissues of stems are derived from the internal L-III. L-I gives rise to the single epidermal layer. L-II gives rise to a single layer of derivatives through several plastochrons (Fig. 13). As further growth occurs, periclinal divisions occur in L-II derivatives so that two or more hypodermal layers of cells may be from L-II. One can see the major contribution of L-III to herbaceous G-W-G stems which are green and to G-G-W stems which are white (Fig. 12) (Stewart, 1965; Stewart et al., 1974). Darker or paler streaks often appear on these stems and in these streaks L-II derivatives have formed cells deeper into the cortex. This increase in tissue from L-II is not uniform around the circumference of the stem (Dermen, 1953, 1960). At some points, L-II derivatives are found as deep as the pericyclic region where adventitious roots are initiated. At different points on the circumference of a 2N-4N-2N chimera, diploid (L-III) and tetraploid (L-II) roots were formed (Dermen and May, 1966).

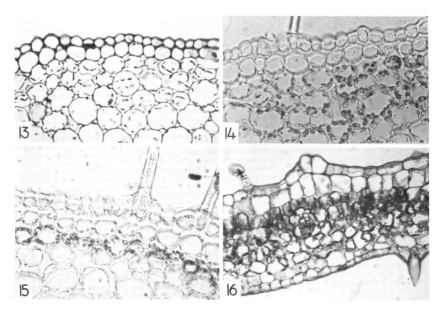

Fig. 13 — Cross section of G-W-G chimeral geranium stem. The subepidermal layer of cells, derived from W L-II of the apical meristem, lacks chloroplastids. Here, as in all epidermal cells except guard cells, the genetically green (G) epidermal cells do not develop chloroplasts. Fig. 14 — Cross section of G-W-G geranium petiole. Only one subepidermal layer of cells is from L-II. Fig. 15 — Cross section of G-G-W geranium petiole. Two subepidermal layers of L-II origin have chloroplasts. Fig. 16 — Cross section near the middle of a G-W-G geranium leaf. The subepidermal layers on both the abaxial and adaxial sides are derived from the W L-II of the apical meristem.

D. Leaves

Formation of a new leaf begins very much like that of a lateral bud. Frequent divisions on the shoulder of the apex result in a protrusion of tissue called the buttress. Although many of the cell divisions in L-I and L-II are apparently periclinal to the surface of the apex, they are anticlinal relative to the surface of the new leaf since derivatives of these two layers usually remain one cell thick throughout most of the leaf. As the leaf axis elongates the hierarchy of layers found in the stem apex is preserved. On the typical shoot with 3 apical layers, derivatives of all 3 are found in the axis. The L-I clone usually produces the epidermis alone although in many linear monocot leaves, a band of internal cells along the margin is derived from L-I. Derivatives of L-II divide anticlinally relative to the new leaf axis and produce a second sheathing layer over the lengthening axis which, as in the stem, largely consists of derivatives of L-III. Sections through petioles and along the midrib or central veins show one or occasionally more hypodermal layers of L-II type (Fig. 14, 15), and the rest L-III. Growth in length of the axis of the leaf differs from that of the stem in that there are no true initial cells. Frequent periclinal and intercalary divisions of distal L-II cells may leave the distal L-III cells far behind the tip of the axis, although L-III usually extends to a point within 2 mm (approximately 200 cells) of the leaf tip. In contrast in the stem tip, the leading L-III cells are just two cell layers behind the distal point. Growth in length of the axis from a very broad buttress essentially completes the growth of linear monocot leaves. It is in these linear monocot leaves that L-I regularly forms marginal internal tissue.

The leaves of G-W-G periclinal chimeras make it obvious that the internal tissue of the blade comes from at least two cell lineages, not from a file or row of submarginal initials as usually depicted in diagrams (Esau, 1977). The typical or average leaf on a G-W-G tobacco or holly shoot has one hypodermal layer of colorless L-II cells over the central green L-III cells on both the abaxial and adaxial sides (Fig. 16, 17) as well as colorless L-II cells making up all the internal tissue around the white margins. Epidermal cells, except guard cells of stomates, do not develop chloroplastids even if gentically green. The margin between the green and white cells of L-II and L-III is very irregular and the width of the white depends upon the number and timing of periclinal and intercalary divisions which occur in L-II (Fig. 6, 19A). In different leaves on the same stem, the amount of tissue from each layer may be quite different. At points along the axis and on different sides of the axis of a single leaf the activity of the layers may also be different (Fig. 18). Most of the

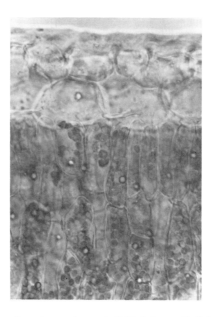

Fig. 17 — Cross section through central area of a G-W-G chimeral holly leaf. The genetically green epidermal cells (L-I) have a slightly greenish cast but no chloroplasts recognizable with a light micro-scope. The hypodermal layer (L-II) is colorless and the internal cells from L-III contain large green chloroplasts.

laminar extension is from intercalary anticlinal divisions. The occasional periclinal intercalary division along the interface of L-II and L-III places a daughter cell in an adjacent layer and subsequent anticlinal intercalary divisions give rise to the isolated darker or lighter areas of green on chimeral leaves (Fig. 6).

The thickness of the extending lamina is determined by the number of cells on the axis participating in the lateral growth. The thickness in terms of number of cell layers is maintained by predominately anticlinal divisions. Daughter cells of periclinal divisions are incorporated into adjacent layers and do not add a new layer of cells (Fig. 7).

E. Reproductive Structures

The development of peduncle and pedicel is like that of stems (Burk et al., 1964; Stewart et al., 1974). The bulk of internal tissue is derived from L-III, the epidermis from L-I, and one, two or occasionally more hypodermal layers are from L-II. Therefore, peduncles and pedicels of

Fig. 18 — Leaf of a G-W-G chimeral banana. The dark areas are derivatives of the G L-III and illustrate its variable contribution to the lamina, and the accommodation by L-II derivatives to produce a leaf of typical size and shape.

G-W-G are white and those of G-G-W are green. Also they grow in length like a stem, i.e., the ultimate source of new cells is from 1 to 3 initial cells in each apical layer. This is evidenced by the dimensions of sectors in the peduncle and pedicel and in their appendages. Sepals have proportions of tissue from the various apical layers similar to those of leaves and thus on G-W-G chimeras the sepals have a white margin and central green area. The bracts of poinsettia show this same pattern in pigment chimeras (Stewart and Arisumi, 1966) as would be expected considering their status as modified leaves. We have found no chimeral plant whose true petals show the pattern of normal and mutant tissue seen in leaves or sepals. This would suggest that all internal tissue of true petals is derived from L-II as has been shown to be the case in cytochimeras of *Datura* (Satina, 1944) and peach (Dermen and Stewart, 1973). The ovary wall of G-W-G chimeras of poinsettia (Stewart, 1965) and geranium (Stewart et al., 1974) are white, indicating a majority of the tissue is derived from L-II. However, cytochimeras of *Datura* (Satina and Blakeslee, 1943) and peach (Dermen, 1960; Dermen and Stewart, 1973) show a major portion of the ovary wall to be of L-III origin. However, both cytochimera (Satina, 1945; Dermen, 1960; Dermen and

Stewart, 1973) and genetic (Baur, 1909; Burk et al., 1964; Stewart, 1965; Stewart and Arisumi, 1966; Kirk and Tilney-Bassett, 1967; Neilson-Jones, 1969; Stewart and Burk, 1970; Stewart et al., 1974) evidence show that both male and female gametes are derived from L-II. The ultimate source of growth of the floral axis is still from one to three initial cells as indicated by the sectors found in flowers and flower parts (Fig. 5).

VI. COMPETITION AND FLEXIBILITY IN ONTOGENY

Competition between cell clones in chimeras operates at several levels. The greatest selective advantage a newly formed lineage can have is one of position. If the event establishing a new clone is in the initial cell of an apical layer with one initial cell, the new clone faces little or no immediate competition. If it is in one of two or three initials, the probability of a shift to a single initial depends upon the effect of the event upon the vigor of the changed initial. As noted earlier, many mericlinal chimeras persist for extensive periods of growth, more than 100 nodes. If the event occurs proximally to the initial cells, the new clone, whatever its vigor, will be left behind unless a lateral bud is formed within the established sector. Sectors from events which occur in differentiating organs are short lived.

To a limited degree there is competition between adjacent cells within a layer at the edge of a mericlinal sector. The same competitive situation faces a cell following its appearance among derivatives of a different layer after a periclinal division. A new clone established by any event within a cell which is not an initial also faces competition for space as a consequence of the extensive intercalary divisions. The size of these late-appearing sectors shows that there is extensive intercalary division during ontogeny of leaf, petal, etc. In the development of various lateral determinant appendages there are no apical or lateral initials or initial layers which are the ultimate source of new cells.

Competition between clones in different layers occurs throughout the growth of the primary body. Again, the most significant arena is at the distal point of the shoot. If an initial cell is replaced or displaced by periclinal division of a cell in an adjacent layer, the source of that cell lineage in all new growth is lost. Replacement or displacement at positions farther from meristematic areas and close to cessation of cell division and differentiation produce smaller and smaller sectors.

We have pointed out (Stewart et al., 1974) that chimeras most useful in development studies are those in which the two cell lineages are of equal vigor. In the geranium chimera carrying the plastid mutant W_1, the amount of leaf tissue from the several apical layers was the same whether they were mutant or wild type (Fig. 19A, B). The irregular edge of the L-II and L-III tissue represented a competitive standoff between the clones. Another plastid mutant, W_2, was apparently less vigorous and at a selective disadvantage. With W_2 L-II and G L-III, the white margins were much narrower than the green margins with G L-II and W_2 L-III (Fig. 19C, D). These margins between L-II and L-III derivatives in the leaf blade, and elsewhere, are determined by the position timing and frequency of periclinal divisions and by the timing frequency and orientation of subsequent anticlinal divisions. These chimeras display a surprising ability to develop a normal primary body through quite different ontogenetic pathways.

There is often a greater stability of the apical layers in elongation of the shoot axis than in the development of the determinant appendages. This has been noted by Clowes (1961), Tilney-Bassett (1963), Dermen (1969), Stewart et al. (1974), and Stewart and Dermen (1974). A G-W-G

Fig. 19A — Leaf on G-W_1-G chimeral geranium shoot. The white tissue carries the plastogene mutant W_1. Fig. 19B—Leaf on a G-G-W_1 chimeral geranium shoot. The white tissue from L-III in this leaf is as extensive as the green tissue from L-III in the G-W_1-G leaf. Fig. 19C — Leaf on a G-W_2-G chimeral geranium plant. W_2 is a different plastogene mutant. The amount of white tissue from W_2 L-II in this leaf is far less than the white tissue from W_1 L-II in the leaf pictured in Fig. 18A. Fig. 19D — Leaf on G-G-W_2 chimeral geranium. There is far less tissue from W_2 L-III in this leaf than from G L-III in the G-W_2-G leaf. However, there is correspondingly more tissue from G L-II and the leaf is of typical size and shape.

chinese privet grown at 16 C produced leaves with a very narrow margin of white tissue compared to plants grown at 26 C. When the 16 C plants were moved to 26 C, growth of the apical and lateral meristems produced leaf patterns typical of 26 C growth (Fig. 11). Conditions which greatly modified the contribution of the apical layers to development of leaves did not affect the hierarchy of layers in the terminal or lateral apical meristems (Stewart and Dermen, 1975).

In chimeras where the vigor of the components is similar, the occurrence of periclinal divisions does not upset the usual pattern of development. For example, the change in number of green cell layers as the result of periclinal division during extension of a leaf blade does not increase the number of cell layers or leaf thickness (Fig. 7). On the other hand, many mutant clones differ markedly in their vigor and abnormal growth and development occurs (Clowes, 1961; Bergann and Bergann, 1962; Tilney-Bassett, 1963; Dermen, 1965; Neilson-Jones, 1969). However, most of the chimeras we have observed display a surprising ability to develop a normal primary body through quite different ontogenetic pathways. For example, the leaves of geranium chimeras carrying the W_2 plastogene were the same size and shape in G-W-G and G-G-W arrangement although the amount of tissue derived from L-II and L-III were extremely different (Stewart et al., 1974). In tobacco, privet and other chimeras (Stewart and Dermen, 1975) successive leaves along a branch may have very little or very extensive contribution from L-III. Along the axis of a leaf, extension into the lamina by L-III may be limited or extensive (Fig. 18).

VII. METABOLIC COOPERATION

Most characters expressed and measured at the cellular level have shown complete independence in plant chimeras. This is true of cell size, chromosome number, plastid development, anthocyanin synthesis, pH, fragrance, etc. Cell size and chromosome number differences in cytochimeras have been observed by many (Satina et al., 1940; Dermen, 1945; Dermen and May, 1966; Stewart et al., 1972). Normal plastid development in cell lineages adjacent to albino cell lineages has been observed many times since Baur (1909) first described periclinal chimeras. Adjacent cell lineages of different genotype have shown complete independence of pigment synthesis (Sagawa and Mehlquist, 1957; Asen et al., 1971; Stewart et al., 1972) although Pereau-Leroy (1974) described an interaction between layers of different genotype in a *Dianthus* chimera.

A graft chimera of *Camellia japonica,* and *C. sasanqua* (Stewart et al., 1972) has shown independence of expression at the cellular level of color, pH, cell size and chromosome number (Table I). The pollen size, pollen shape, epidermal hairs, and fragrance were characteristic of the component which formed them. There was interaction or cooperation with respect to a number of characters expressed at the organ or whole plant level. The flowering season of the chimera spanned that of both components and the number of petals was intermediate. Most striking was the formation of ovary, anthers and gametes by the *C. japonica* L-II and L-III component of the chimera (Fig. 20A-H). When isolated from the chimera, the *C. japonica* was completely sterile, i.e., the double flowers formed only petals, no gynoecium or androecium. Within the *C. sasanqua* epidermal tissue complete reproductive structures were formed which contained ovules and pollen grains which germinated.

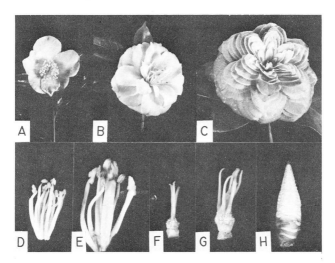

Fig. 20 — Flowers and reproductive structures from the graft chimera *Camellia* + 'Daisy Eagleson' (DE) and the component species isolated from it. Fig. 20A — Flower of hexaploid *C. sasanqua* (SAS) isolated from DE. Fig. 20B — Flower of DE which has L-I of SAS and L-II and L-III from *C. japonica* (JAP). Fig. 20C — Flower of diploid JAP isolated from DE. Fig. 20D — Stamens from SAS. Fig. 20E — Stamens from DE. JAP produced no stamens. Fig. 20E — Receptacle and gynoecium SAS. Fig. 20F — Receptacle and gynoecium of DE. JAP produced no gynoecium. Fig. 20G — The large receptacle of JAP showing many petal-base scars and lack of gynoecium at the distal end.

VIII. DETERMINATIVE EVENTS AND CELL FATE

A. *The Change from Unorganized to Organized Growth*

The first determinative event is the change from unorganized to organized growth. This occurs when random orientation of cell division becomes controlled in such a way as to form and perpetuate an apical meristem. Classical plant embryologists felt that the very early divisions of the zygote precisely determined the developmental fate of the cells formed thereafter. Recent work has questioned this concept and has shown that there are many exceptions to the classical patterns

TABLE I.

Characteristics of a Periclinal Chimera (L—I=C. sasanqua, L-II & III= C. japonica) and the Species Reisolated from it.

Character	*C. sasanqua*	chimera	*C. japonica*
FLOWER			
Type	Single	Incomplete double	Complete double
Diameter (cm)	ca. 5.5	ca. 7.3	ca. 9.0
Color[a]	Light pink	Light pink	Deep purplish pink
Fragrance	Musty apple	Musty apple	None
GYNOECIUM			
Ovary	Present	Present	Absent
	Pubescent	Pubescent	(b)
Ovules	Present	Present	Absent
Style	3-fid	6-8-fid	Absent[c]
ANDROECIUM			
Number of anthers	60-70	20-45	Absent
Pollen size (μ)	78	90	Absent
Pollen shape	Round	Angular	Absent[d]
Filaments	Free to base	Connate at base	Absent[e]
Meiotic metaphase	Many multivalents	15 bivalents	Absent[f]
PETAL			
λ max.[g] epidermal cells	538 nm	538 nm	532 nm
λ max. internal cells	538 nm	532 nm	532 nm
pH epidermal cells	4.9	4.9	3.5
pH whole petal	4.8	4.3	3.5
Number of petals	5-7	25-50	60-80
ROOT			
Chromosome number	ca. 90	30	30
SHOOT			
Leaf, stem, bud	Pubescent	Pubescent	Glabrous
Cell size in apex	L-I Large	L-I Large	L-I Small
	L-II Large	L-II Small	L-II Small
	L-III Large	L-III Small	L-III Small

a. Color notations according to Kelly, 1965.
b. *C. japonica* typically glabrous.
c. *C. japonica* typically 3-fid.
d. *C. japonica* typically angular.
e. *C. japonica* typically connate
f. *C. japonica* typically has 15 bivalents.
g. Wavelength of maximum absorption.

of cell division and that cell divisions are at random in early globular stage of embryogenesis (Jensen, 1964, 1976). Random position and orientation of mitoses are also characteristic of the unorganized growth of callus, either in the petri dish or in the plant.

Our work with chimeras suggests that this first determinative event involves a decision or response of a group of cells which become the initial cells of the new apical meristem. The growth of the plant primary body is from a layered apical meristem in which cell divisions are quite precisely oriented and coordinated. The initiation of such ordered division within the early embryo or callus must depend, at least in part, upon the occurrence of the critical arrangement of a group of cells in the form of an apical meristem which then establishes gradients of growth regulating substances to perpetuate ordered divisions. Since angiosperms typically have three layered apical meristems, each layer having from one to three initial cells, the formation of a new apical meristem (adventitious bud) must involve from three to nine cells. However, the origin of adventitious buds and shoots has generally been thought to be from a single cell (Sparrow et al., 1960; Burk, 1975; D'Amato, 1977). Principal evidence for single cell origin has been the same chromosome number in root and shoot and in different flowering branches. However, both the roots and branches are derivatives of the regenerated shoot's adventive apical meristem and their similarity is not proof of single cell origin. Moreover, the literature has reference to many events which demonstrate the origin of organized growth from more than one cell. The *Solanum* graft chimeras of Winkler (1907) had to originate from at least two cells as did all the woody graft chimeras discussed by Nielsen-Jones (1969). Emsweller and Brierley (1940) obtained cytochimeras in the adventitious shoots regenerated on colchicine treated scales of *Lilium*. Clowes and Juniper (1968) pictured a chimeral adventitious bud on a leaf of variegated *Sedum*. Arisumi and Frazier (1968) reported one cytochimeral adventitious shoot out of 28 after colchicine treatment of *Saintpaulia* leaf petioles. Carlson and Chaleff (1975) obtained chimeral shoots from mixed callus. Stewart and Dermen (1970b) obtained 80 adventitious shoots from chimeral cultivars of *Chrysanthemum*. Twenty-seven were chimeral, 20 were not chimeral, and 33 could not be determined without further analysis. One of the adventitious chimeral chrysanthemum shoots had an arrangement of layers which proved it came from at least 3 cells. A similar chimeral shoot from 3 cells occurred in a propagation of a variegated *Sansevieria* in our greenhouse. We suggest that the initiation of an apical meristem always involves or requires a group of cells. These may have all been *ultimately* derived from a single cell and therefore the new apical meristem may be homogeneous, but at the time of determination, a minimum of 3 cells is involved. Davidson et al. (1976) questioned whether a disorganized *group* of cells could become ordered with respect to each other and thereby produce recognizable structures in contrast to the development of such structures from a single cell in ordered sequence. We feel that the *only* way ordered growth is initiated is by a *group* of cells. This has been demonstrated in regeneration from callus and is suggested in embryogenesis. The origin of the apical meristem in embryo formation could be investigated using the chimeral zygotes produced in the semigametic cotton lines of Turcotte and Feaster (1967). The fact that chimeral cotton seedlings are obtained with haploid tissue of maternal and paternal origin proves that both products of the first division of the zygote can be involved in embryo formation. One daughter is not always relegated to the suspensor. Another approach to studying determinative events in embryogeny would be with hetero-plastidic zygotes obtained by male transmission of plastogenes or by fertilization of heteroplastidic eggs. Early segregation to homo-plastidic clones could yield information on the origin of the apical meristem of the embryo.

Davidson et al. (1976) suggests that the emergence of organized structures from disordered callus results from divisions in which products remain in close contact enabling an intimate local conversation or exchange of substances to become established and maintained. Sub-sequent regulated divisions lead to the formation of organized struc-tures. Their concept agrees well with our evidence that a group of cells is involved in initiation of ordered growth. Once established the order is self perpetuating as indicated by the continued growth as a shoot of isolated apical domes (Smith and Murashige, 1970). D'Amato (1977) points out that apex cultures, in contrast to callus, give cytogenetic stability. Regeneration from callus in culture is influenced by exogenous growth regulators (Davidson et al., 1976) but continued organized growth is independent. A similar hormonal influence is suggested by the fact that chrysanthemum stems (Stewart and Dermen, 1970b) formed adventitious buds only in callus immediately adjacent to the scar of excised leaves and lateral buds.

B. *Ultimate Fate*

Once the axis of the plant with its apical meristems is established, the development and differentiation of the primary body is very flexible, at least in terms of the ultimate fate of any cell or clone. Jensen (1976) suggests that ". . . the position of a cell in an embryo and its relationship to centers of hormone production, will determine its pattern of differentiation." Steward (1964) in discussing embryogenesis from somatic cells points out that while any carrot cell may be totipotent, once they are included in an apical meristem "what each cell does, . . . , is as much a function of *where it is* as of *what it is*." In the distal position of the shoot apical meristem an initial cell gives rise to a continuing sector of daughter cells added to the primary body. If not in the distal position a cell in the meristem gives rise to shorter and narrower sectors and is left behind. If in the right position on the shoulder of the apex, a clone will participate in leaf and lateral bud formation. During development of a leaf the contribution of a clone to petiole or blade may be extensive or limited and is accommodated by flexibility in contribution of the other clones involved. A particular cell in a developing leaf may divide only a few times while the sister cell divides many times as evidenced by twin spots of very different size. In any position a cell may divide anticlinally or periclinally. Within the anticlinal plane the orientation and sequence of divisions may be random, producing wide sectors, or ordered, producing narrow streaks. Similar flexibility is seen in the ontogeny of the stem and reproductive structures. This extreme variability in direction and frequency of cell division is under overriding genetic control which results in final normal or typical morphology.

Our chimeral studies have led us to the conclusion that the ultimate fate of a given cell may not be determined until the final division, and that the position occupied at that time determines final form, subject to limitation of its genetic potential. In a G-W-W chimera the single green palisade cell near the edge of the white leaf was destined to be a colorless epidermal cell until a final intercalary periclinal division placed it in the hypodermal layer where its potential for chlorophyll synthesis was expressed. Similarly in our G-W-W and W-W-G tobacco chimeras (Stewart and Burk, 1970) a cell apparently destined to be in the epidermis or ovule wall became a megagametophyte. Vascular bundles develop along a path dictated by spatial and physiological gradients, not along paths laid down by predetermined cell lineages.

Similar events in many different chimeral plants have clearly demonstrated the amazing flexibility in the ontogenetic pathway to final form. The flexibility is possible because cell fate is not determined early in development and sister cells may mature or differentiate to diverse form and function.

ACKNOWLEDGMENT

I wish to record my debt to the late Dr. Haig Dermen for much fruitful discussion, advice, and encouragement which continued from his retirement in 1965 until his death on July 31, 1977.

REFERENCES

Arisumi, T., and Frazier, L. C. (1968). *Proc. Am. Soc. Hort. Sci.* **93,** 679-685.

Asen, S., Stewart, R. N. and Norris, K. H. (1971). *Phytochemistry* **10,** 171-175.

Bain, H. F. and Dermen, H. (1944). *Am. J. Bot.* **31,** 581-587.

Barrow, J. R. and Dunford, M. P. (1974). *J. Hered.* **65,** 3-7.

Barrow, J. R., Chaudhari, H. K. and Dunford, M. P. (1973). *J. Hered.* **64,** 222-226.

Baur, E. (1909). *Z. Indukt. Abstammungs-Vererbungsl.* **1,** 330-351.

Bergann, F. and Bergann, L. (1962). *Der Zuchter* **32,** 110-119.

Boke, N. H. (1948). *Am. J. Bot.* **35,** 413-423.

Burk, L. G. (1975). *Plant Sci. Ltrs.* **4,** 149-154.

Burk, L. G., Stewart, R. N. and Dermen, H. (1964). *Am. J. Bot.* **51,** 713-724.

Buvat, R. (1955). *Ann. Biol.* **31,** 595-656.

Carlson, P. S. and Chaleff, R. S. (1975). *In:* "Genetic Manipulation with Plant Material" (L. Ledoux, ed.), Vol. 3, pp. 245-261. Plenum Press, New York.

Clowes, F. A. L. (1957). *Heredity* **11,** 141-148.

Clowes, F. A. L. (1961). "Apical Meristems". Blackwell. Oxford.

Clowes, F. A. L. and Juniper, B. E. (1968). "Plant Cells". F. A. Davis Co., Philadelphia.

D'Amato, F. (1977). *In:* "Plant Cell, Tissue, and Organ Culture" (J. Reinert and Y. P. S. Bajaj, eds.), pp. 343-357. Springer-Verlag, Berlin.

D'Amato, F. and Avanzi, S. (1968). *Caryologia* **21,** 83-89.

Davidson, A. W., Aitchson, P. A. and Yeoman, M. M. (1976). *In:* "Cell Division in Higher Plants" (M. M. Yeoman, ed.), pp. 407-432. Academic Press, New York.

Dermen, H. (1945). *Am. J. Bot.* **32,** 387-394.

Dermen, H. (1947). *Proc. Am. Soc. Hort. Sci.* **50,** 51-73.

Dermen, H. (1951). *Am. J. Bot.* **38,** 753-760.

Dermen, H. (1953). *Am. J. Bot.* **40,** 154-168.

Dermen, H. (1960). *Am. Hort. Mag.* **39,** 123-173.

Dermen, H. (1965). *Am. J. Bot.* **52,** 353-359.

Dermen, H. (1969). *Cytologia* **34,** 541-558.

Dermen, H. and May, C. (1966). *Forest Science* **12,** 140-146.

Dermen, H. and Stewart, R. N. (1973). *Am. J. Bot.* **60,** 283-291.

Dommergues, R. (1962). *Third Congress of the European Assoc. for Research on Plant Breeding,* p. 115-139.

Emsweller, S. L. and Brierley, P. (1940). *J. Hered.* **31,** 223-230.

Esau, K. (1965). "Plant Anatomy". John Wiley & Sons, Inc., New York.

Esau, K. (1977). "Anatomy of Seed Plants". John Wiley & Sons, Inc., New York.

Esau, K., Foster, A. S. and Gifford, E. M., Jr. (1954). *VIII Int. Congr. Bot.*, Paris, Rep. and Comm., Sec. 7-8.

Farestveit, B. (1969). *Arrsskr. k. Vet. Landbohojsk.*, p. 19-33.

Gifford, E. M. and Corson, Jr., G. E. (1971). *Bot. Rev.* **37**, 143-229.

Gregory, R. P. (1915). *J. Genet.* **4**, 305-322.

Jensen, W. A. (1964). *In:* "Meristems and Differentiation". *Brookhaven Symp. Biol.* **16**, 179-202.

Jensen, W. A. (1976). *In:* "Cell Division in Higher Plants" (M. M. Yeoman, ed.), pp. 391-405. Academic Press, New York.

Jones, D. F. (1937). *Genetics* **22**, 484-522.

Jorgensen, C. A., and Crane, M. B. (1927). *J. Genet.* **18**, 247-273.

Kelly, K. L. (1965). *Color Engng.* **3**, 2.

Kirk, J. T. O. and Tilney-Bassett, R. A. E. (1967). "The Plastids". W. H. Freeman and Company, London.

Lange, F. (1927). *Planta* **3**, 181-187.

Lyndon, R. F. (1976). *In:* "Cell Division in Higher Plants" (M. M. Yeoman, ed.), pp. 285-314. Academic Press, New York.

Michaelis, P. (1958). *Proc. 10th Intern. Congr. Genet.* **1**, 375-385.

Nougarede, A. (1967). *In:* "Int. Rev. Cytol." (G. H. Bourne and J. F. Danielli, eds.) **21**, 203-351. Academic Press, New York.

Neilson-Jones, W. (1969). "Plant Chimeras". Methuen & Co., Ltd., London.

Newman, I. V. (1965). *J. Linn. Soc. London Bot.* **59**, 185-214.

Pereau-Leroy, P. (1974). *Radiation Bot.* **14**, 109-116.

Popham, R. A. (1958). *Am. J. Bot.* **45**, 198-206.

Pratt, C., Ourecky, D. K. and Einset, J. (1967). *Am. J. Bot.* **54**, 1295-1301.

Ramulu, K. S., Devreux, M., Ancora, G. and Laneri, U. (1976). *Z. Pflanzenzeucht.* **76**, 299-319.

Sagawa, Y. and Mehlquist, G. A. L. (1957). *Am. J. Bot.* **44**, 397-403.

Savelkoul, R. M. H. (1957). *Am. J. Bot.* **44**, 311-317.

Satina, S. (1944). *Am. J. Bot.* **31**, 493-502.

Satina, S. (1945). *Am. J. Bot.* **32**, 72-81.

Satina, S. and Blakeslee, A. F. (1941). *Am. J. Bot.* **28**, 862-871.

Satina, S. and Blakeslee, A. F. (1943). *Am. J. Bot.* **30**, 453, 462.

Satina, S., Blakeslee, A. F. and Avery, A. G. (1940). *Am. J. Bot.* **27**, 895-905.

Smith, R. H. and Murashige, T. (1970). *Am. J. Bot.* **57**, 562-568.

Soma, K. and Ball, E. (1964). *In:* "Meristems and Differentiation". *Brookhaven Symp. Biol.* **16**, 14-45.

Sparrow, A. H., Sparrow, R. C. and Schavier, L. A. (1960). *African Violet Mag.* **13**, 32-37.

Steward, F. C. (1964). *In:* "Meristems and Differentiation". *Brookhaven Symp. Biol.* **16**, 73-88.

Stewart, R. N. (1965). *Genetics* **52**, 925-947.

Stewart, R. N. and Arisumi, T. (1966). *J. Hered.* **57**, 216-220.

Stewart, R. N. and Burk, L. G. (1970). *Am. J. Bot.* **57**, 1010-1016.

Stewart, R. N. and Dermen, H. (1970a). *Am. J. Bot.* **57**, 816-826.

Stewart, R. N. and Dermen, H. (1970b). *Am. J. Bot.* **57**, 1061-1071.

Stewart, R. N. and Dermen, H. (1975). *Am. J. Bot.* **62**, 935-947.

Stewart, R. N., Meyer, F. G. and Dermen, H. (1972). *Am. J. Bot.* **59**, 515-524.

Stewart, R. N., Semeniuk, Pete and Dermen, H. (1974). *Am. J. Bot.* **61**, 54-67.

Thompson, M. M. and Olmo, H. P. (1963). *Am. J. Bot.* **50**, 901-906.

Tilney-Bassett, R. A. E. (1963). *Heredity* **18**, 265-285.

Turcotte, E. L. and Feaster, C. F. (1967). *J. Hered.* **58**, 55-57.

Vig, B. K. (1973). *Genetics* **73**, 583-596.

Wardlow, C. W. (1968). "Morphogenesis in Plants: A Contemporary Study." Methuen, London.

Woods, M. W. and DuBuy, H. G. (1951). *J. Natl. Cancer Inst.* **11,** 1105-1151.
Winkler, H. (1907). *Ber. Dtsch. Bot. Ges.* **25,** 568-576.

The Development of Spacing Patterns
in the Leaf Epidermis

Tsvi Sachs

Department of Botany
The Hebrew University
Jerusalem, Israel

I. THE PROBLEM

The cellular structures of the epidermis of leaves of *Sedum nicaeense* All. and *Pisum sativum* L. are illustrated in Figs. 1 and 2. The major point relevant to the study presented here is that there are two types of cells: guard cells forming stomata and other epidermal cells. The distances between the stomata vary greatly, but they appear to be more uniform than would be expected if distribution were random. The leaf is therefore a relatively simple example of orderly, or patterned differentiation. The type of pattern found in the epidermis was called a "spacing pattern" by R. Gordon (Wolpert, 1971).

Patterned differentiation is an essential biological phenomenon but is very poorly understood. The leaf epidermis has some basic advantages for the study of spacing phenomena. The pattern involved is relatively simple and its formation does not involve any cell movement. Differentiation occurs on the surface and late in leaf development, so

161

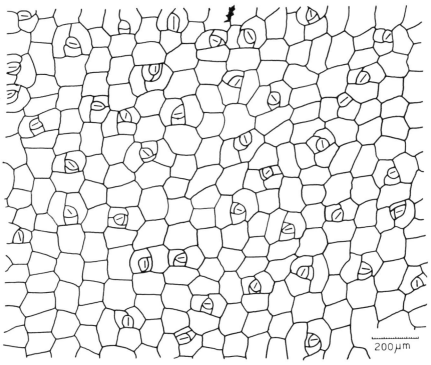

Fig. 1. Exact drawing of the cells of the lower epidermis of the center of a leaf of *Sedum nicaeense* All., showing the pattern of stomata spacing. Prepared from a nail-polish replica of the surface of a leaf which had been washed with ethanol. Drawn by placing transparent paper on a screen of a Reichert "Visopan" projection microscope.

that it can be observed and manipulated. And, finally, many examples of various complexity are available in the plant kingdom.

The spacing patterns of the stomata raise the general question of the types of processes which control the orderly aspects of this differentiation. This question assumes that the meristematic cells which form the epidermis are able to undergo differentiation to at least two types of cells, and emphasizes the problem of the nature of the processes which cause this differentiation to be orderly. The question might be clarified by a consideration of the possible answers, or the possible factors which could contribute to this order:

(a) Correlative inhibition. A stoma may influence the differentiation of the surrounding tissues, inhibiting the formation of additional stomata. If this inhibitory effect decreased with the distance from its source a spacing pattern would be formed (Bunning and Sagromsky, 1948; Bunning, 1948, 1965). Similar suggestions have been made

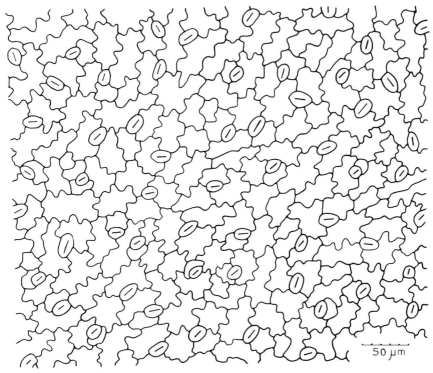

Fig. 2. Exact drawing of the lower epidermis of a pea leaf showing the pattern of stomata spacing. Prepared as was Fig. 1, using a stipule on the 7th node (from the root) of *Pisum sativum* L. cv Alaska.

concerning other biological spacing patterns, such as the distribution of hairs on insects (Wigglesworth, 1940). The mechanism by which inhibition could be exerted will be discussed below; it is not relevant at this stage in which general factors are defined.

(b) Cell lineage or clone formation. The order and orientation of the cell divisions by which a meristematic cell gives rise to both stomata and epidermal cells could be the basis for spacing patterns (Bunning and Sagromsky, 1948; Bunning, 1965; Korn, 1972). It is obvious how this spacing pattern would arise if each stoma was formed surrounded by epidermal cells, so that any two stomata would be necessarily separated, or spaced. Orderly spacing, however, can also be expected if each stoma has epidermal cells on one side, provided that the orientation of these stoma-epidermal cell complexes was the same for large regions of the leaf (Sachs, 1974). Bunning and Sagromsky (1948) stated that cell lineage could only be important in the determination of spacing when stomata formation was synchronous. As will be shown below, however, this is not logically correct: the relative orientation of the stomata-epidermal cell complexes is important, but not the relative time of their formation.

(c) Patterned influences from outside the epidermis. The spacing pattern may reflect inductive influences of another pattern close to it. In the case of the epidermis, such control could be exerted by the minor leaf veins or by the pattern of the cells in the mesophyll. No known environmental influences have a spacing pattern where the units are less than 100μm distant from one another.

These three factors or mechanisms can be worded so as to include all possible controls. This is so because they are basically the anwers to two questions: are the controls of the spacing within the epidermis or external to it and, if internal, are they dependent on intercellular interactions or on intracellular mechanisms. It should be noted, however, that these three possible mechanisms do not exclude one another. The central question of this study, therefore, is not which of these factors is correct but what is their relative role in the formation of the spacing pattern. The idea of a correlative inhibition exerted by developing stomata is often considered to be "obviously" correct (Sinnott, 1963; Kuhn, 1965; Lang, 1973). This mechanism can certainly work, as has been demonstrated by computer studies (Gierer and Meinhardt, 1972; Korn, 1972; Korn and Fredrick, 1973). The evidence for the role of this mechanism in the formation of spacing patterns of stomata, however, is at best meagre and the idea has probably been so widely accepted only because it is simple and attractive.

II. THE MEASUREMENT OF STOMATA SPACING

So as to understand the formation of a spacing pattern we must start by measuring or analysing its parameters. Even the basic idea that the pattern is not random is not obvious: Bunning and Sagromsky (1948) state as evidence for orderliness the observation that no two stomata touch one another. This, however, is not strictly correct, since such stomata can be found though their frequency is low (for example, see Gopal and Shah, 1970).

Methods for the measurement of the degree of orderliness of spacing patterns or, in other words, their deviation from a random distribution have been developed in connection with ecological problems. The distribution of plants or animals in a field is often a problem of a spacing pattern, even though the scale is so totally different from the one we are dealing with here. The method which has been most widely used for the study of spacing patterns in tissues is the parameter "R" (Clark and Evans, 1954). This parameter is the ratio between the measured average distances between nearest neighbors and the corresponding distances predicted for a random distribution of the same density. Any significant deviation from R = 1, therefore, indicates that the pattern deviates from the random distribution. This method has been used for stomata (Korn and Fredrick, 1973), for the pattern of hairs in sheep (Claxton, 1964) and in insects (Lawrence and Hayward, 1971). In the cases studied here, R of stomata distribution in *Sedum* was 1.4±0.1 and in *Pisum* 1.5±0.1. These measurements show that the pattern is more regular, or spaced, than would be expected for a random distribution but is far from perfect order, in which case R would be somewhat over 2 (Clark and Evans, 1954). This conclusion, however, does not lead us further than the intuitive expectation from the simple observation of stomata patterns.

So as to characterize the spacing pattern of stomata we must know how the presence of a stoma is correlated with the events in the surrounding tissue. Curves which describe various theoretical correlations of this type (known as radial distribution functions in physics) are illustrated in Fig. 3. In terms of the various factors

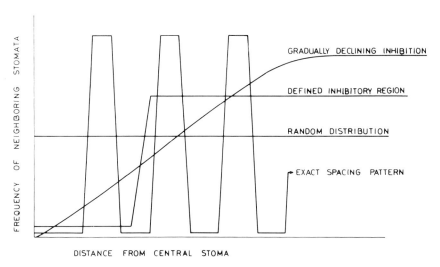

Fig. 3. Theoretical curves showing possible average relative frequencies of stomata as a function of the distance from any randomly selected stoma. A horizontal line shows a lack of any correlation between the presence of a stoma and the occurrence of additional stomata in its vicinity. Note that various deviations from such a horizontal line are possible and that they would be meaningful concerning the nature of the spacing pattern studied.

considered above as possibly contributing to the formation of stomata spacing patterns these correlations could have specific meaning. If we assume an inhibitory effect of a stoma on its surroundings, the question would be how far from the stoma does this inhibition extend and whether it changes gradually or abruptly as the distance from the stomata increases. If we assume clonal development as the basis of pattern formation, the question would be to what distance from a stoma does a clone extend, as expressed by the absence of additional stomata. If we assume an influence of other tissues, the question is what are the parameters of the inducing pattern for which one must look. The relation between the frequency of neighboring stomata and the distance from any given stoma could therefore tell us something concrete concerning the basic parameters of stomata spacing patterns. Similar curves have been used by Korn (1972).

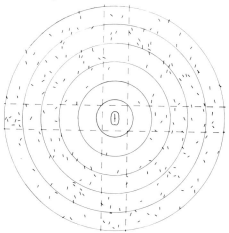

Fig. 4. Explanation of the method used to construct the curves showing the relation of distance from any randomly selected stoma and the average frequency of additional stomata. A Reichert "Visopan" projection microscope was used to obtain an enlarged view of a nail polish copy of the surface of an epidermis, but pictures would do just as well. Transparent paper was placed on the microscope screen and the stoma found by chance to be closest to the center was moved to the actual center. The paper was oriented so that the axis of the stoma was along the axis of the stoma drawn at the center of the paper. The pores of all additional stomata in the field observed were marked on the paper by a short line. This process was repeated at least 25 times for different microscopic fields, using the same transparent paper. The drawing of the figure shows the actual results when the epidermis of a pea stipule was observed.

Curves were constructed on the basis of at least 8 cases of the type shown in the figure, which included at least 200 microscopic fields. The stomata pores were counted in groups according to their distance from the center with the aid of lines such as those which appear in the figure. These counts were first made separately for the parallel and perpendicular directions relative to the axis of the central stomata. When no significant differences were found between these directions, as was the case for the plants used in this study, counts were also made for rings surrounding the center. The numbers were corrected so as to represent numbers of stomata in an arbitrary unit area, cancelling the effect of increased ring size as the distance from the center increases. For the expected meaning of such results, see Fig. 3.

The next question is how to construct such curves describing the correlation between a stoma and the events in the surrounding tissues. The method I have used is explained by Fig. 4. The essential point about this method is that many fields surrounding many stomata were superimposed upon one another. The stomata placed in the center of the fields studied were randomly chosen, so that the measurements represent the properties of the entire spacing pattern and chance events next to individual stomata should be cancelled out.

Fig. 5 is a curve constructed in this way for the stomata of the lower epidermis of *Sedum*. With few exceptions, stomata are absent from the immediate neighborhood of any chosen stoma, up to a distance of 100μm. There is, therefore, a fairly clear "stomata-free" region surrounding each stoma. This "stomata-free" region disappears completely at about 180μm and beyond it there is no correlation between

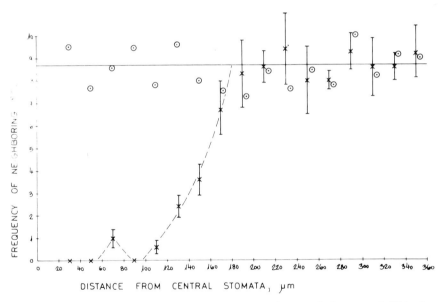

Fig. 5. The relation of the distance from any randomly selected stoma to the frequency of additional stomata in the epidermis of *Sedum*. The straight line at a frequency of 8.7 stomata per arbitrary unit area represents the result calculated from stomata density when a random distribution was assumed; in this case, stomata frequency would not change as a function of the distance from a given stoma. The circles are the results of counts made when the center of the field was a random point, before a nearby stoma was moved to this center (see Fig. 4). As expected, these points do not deviate significantly from the calculated straight lines and they serve as a measure of the variability of the results. The X signs mark the results of counts made when a stoma was at the center; note that with few exceptions there is a clear "stomata-free" region surrounding the central stomata. This region is the only significant deviation of the counts from the horizontal, calculated line at which there is no correlation between the distance and the occurrence of additional stomata.

the presence of a central stoma and the frequency of additional stomata. Fig. 6 is the corresponding curve for pea epidermis. The same conclusion is suggested by this curve as well: that the only orderly aspect of the distribution of the stomata is the presence of a fairly well defined "stomata-free" region surrounding each stoma. The same result has also been obtained for a number of other plants (Sachs, 1974; Marx and Sachs, 1977; and unpublished results).

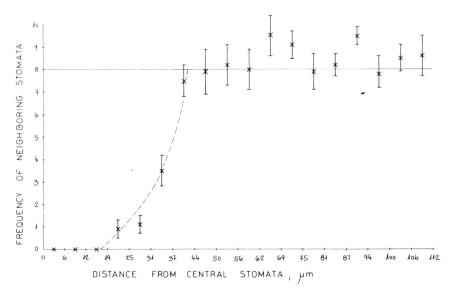

Fig. 6. The relation of the distance from any randomly selected stoma and the frequency of additional stomata in the lower epidermis of a pea stipule. This curve was prepared in the same way as Fig. 5 (see Fig. 4 for explanation). Note that the only deviation of the counts from a straight horizontal line is the presence of a "stomata-free" region next to the central stomata. This "stomata-free" region ends at about 40μm from the stomata, and beyond this distance there is no correlation between stomata frequency and the presence of stomata in the center of the field.

These curves thus lead to a clear conclusion: whatever the factors which determine stomata spacing, they must act only in preventing the appearance of stomata close together, up to a fairly well defined and measurable distance. Beyond these distances of "stomata-free" regions there is no partial inhibition of any other effect correlated with the presence of the neighboring stoma. The distances involved correspond to the size of one or, at most, a few epidermal cells.

III. THE DEVELOPMENTAL ORIGIN OF STOMATA PATTERNS IN SEDUM

The next question is the role of the various possible factors mentioned above in the formation of the "stomata-free" regions which are the basis of stomata spacing patterns. The first to be considered here will be "cell lineage" or clone formation in *Sedum*. The reason for this choice is that cell lineages of stomata development can be readily followed, and the size of the "stomata-free" region predicted from such developmental studies can be compared with that measured by Fig. 5 (Sachs, 1974). *Sedum* was chosen because, as will be seen below, the results in this case are relatively simple to interpret. *Pisum* represents a more complicated case.

Stomata development has been studied since the last century (for example, see de Bary, 1877) and in very many plants. A developmental sequence has always been constructed from the observation of young stages of different stomata. There is no need to use many leaves for this purpose: different developmental stages can generally be found next to one another and in the same young leaf the basal regions are generally at a younger developmental stage than those closer to the tip.

Fig. 7 illustrates young stages of the development of stomata of *Sedum nicaeense*. The earliest stages (marked by the letter "a") are

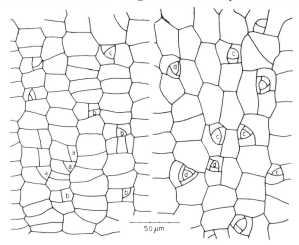

Fig. 7. Two young regions of the epidermis of *Sedum*, drawn from the base of immature leaves. Early stages of stomata development can be recognized by the presence of unequal divisions. The letters mark the sequence of these stages as it may be assumed on the basis of the number of unequal divisions. Note that stomata formation often starts in neighboring cells, that it forms epidermal cells as well as stomata and that it does not follow one exact sequence in terms of the orientation and location of the unequal divisions.

characterized by the presence of an unequal division. The smaller of the two products of an unequal division divides unequally, and this process is repeated a number of times. Finally, stomata are formed by an equal division (Fig. 8) of a small cell; this stoma is surrounded by epidermal cells originating from the same mother cell which first divided unequally.

Fig. 8 illustrates the relation between the "stomata-free" region predicted from this developmental sequence and the same region as measured for the mature epidermis by the curve of Fig. 5. The cells which can be assumed to have originated from one mother cell are indicated by parallel lines of one orientation. The minimal and maximal limits of the "stomata-free" region expected for the central stoma are indicated by circles. It can be seen that the correspondence between the minimal distances which can be predicted from development and those measured (Fig. 5) is remarkably good. The essential point is that stomata that are at distances at which the frequency of their occurrence was measured to be quite independent of one another (180μm) can arise from neighboring mother cells.

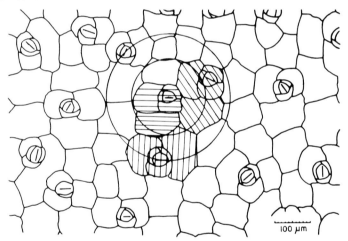

Fig. 8. An enlarged region of mature *Sedum* epidermis. The cells which are assumed to have originated from the same original unequal division of one mother cell (see Fig. 7) are marked in 3 cases by parallel lines of a given direction. The circles which surround the central stoma show distances of 100 and 180μm; these are the minimal and maximal limits of the "stomata-free" region measured by the results plotted in Fig. 5. Note that a "stomata-free" region of the size which was measured could arise from the developmental sequence leading to stomata formation. Thus stomata at distances at which their occurrence was not correlated with one another can form from unequal divisions of neighboring cells.

It can be concluded, therefore, that clonal development can account for the distances between stomata which were measured in the case of *Sedum*. Thus clonal development can be the sole basis for the spacing pattern observed in the mature epidermis, even though the development of the stomata is not synchronous (Sachs, 1974; Marx and Sachs, 1977). There is no evidence or reason to assume any influence, inhibitory or otherwise, of a stoma on the events in neighboring cells. Nor is there reason to search for a basis for the distance between stomata in the internal tissues of the leaf, except to the extent stoma development may be correlated to events in these tissues (see below).

IV. PATTERN DEVELOPMENT IN PISUM

The conclusion that clonal development can be the major basis for stomata spacing in *Sedum* raises the question whether it applies to other plants or is a relatively special case. A major complication which can be expected in applying the same reasoning to other plants is that there are generally fewer unequal divisions leading to the formation of a stoma than in *Sedum*. The epidermal cells formed from the same mother cell as a stoma are, therefore, found only on one of its sides and do not surround a stoma as they do in *Sedum*. A developmental sequence in which the stoma is not surrounded by cells of the same origin could certainly produce a spacing pattern, but only under certain conditions. The orientation of the stomata relative to the epidermal cells formed with them must be the same in any given region, so that the stomata are always separated by epidermal cells and thus by a minimal, "stomata-free" distance. In other words, the developmental processes have to have the same polarity, as they in fact do in many Monocotyledons (Sachs, 1974).

In plants with broad leaves, however, such polarity can not be expected for large regions of the leaf if one judges from the orientation of the veins and the course of leaf development. An example of such a leaf is that of *Pisum sativum*, and an enlarged region of its epidermis is shown in Fig. 9. No clear relation between the stomata and the epidermal cells can be found. Furthermore, the number of epidermal cells in contact with a stoma varies, and though there is a small epidermal cell next to many stomata, the orientation of this stoma-epidermal cell complex is not fixed. A clear example of opposite orientations of stomata and small epidermal cells may be seen in the neighboring stomata numbers 9 and 10.

It follows that so as to understand the development of the spacing pattern in peas it is essential to follow the formation of the distance between neighboring stomata and not only the development of individual stomata. This can not be done by reconstructing stomata development from individual stages; it requires the direct observation of the development of a specific region of the epidermis. To the best of my

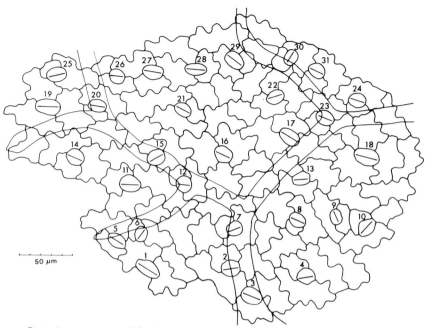

Fig. 9. A mature region of the lower epidermis of a stipule of a pea plant. The development of this region was followed in detail and the numbers identify the stomata in this study (Figs. 10, 11 and 13). The thin lines indicate the limits of the minor veins. Drawn from material cleared with alcohol with a Reichert microscope and drawing apparatus.

knowledge, such a developmental history has not been published. Yet it can be expected that the development of a specific region could be followed since these are surface phenomena which occur when the leaves are fairly large. The major problem for direct observation turned out to be that the surface of the leaf is not flat and changes constantly. Furthermore, the young tissues must be assumed to be delicate and the problem is how to observe them without changing the course of their development.

I have now followed the development of living epidermis using two different methods: daily gelatin prints of specific regions of the surface of a leaf and repeated direct observations using an epi-illumination microscope. Prints were obtained by allowing semi-liquid gelatin to form a solid gel while it was on the surface of the leaf at room temperature. The process of obtaining these prints, however, resulted in some artifacts. The epidermis which matured had a relatively high stomata density and many immature stomata. Therefore, though much more work has been done using the gelatin print method, the results presented below are all based on direct observation using an epi-illumination microscope, a method which caused no detectable artifacts. The conclusions, however, are supported by results obtained using both methods.

Direct observation was carried out with a Wild epi-illumination microscope to which a Reichert drawing apparatus was attached. Drawing rather than photography was used because the surfaces observed were often not flat and constant focusing was required. So as to adjust and move the plant under the microscope it was tied to a simple Narishige micromanipulator while it was being observed. Since pea is a climber, it was possible to move the young leaves without moving the roots or damaging the plant. The area studied was recognized for each observation with the aid of black marks placed some distance from it and by the local configuration of the veins, cell walls and stomata. To avoid the possibility of local heating due to light, the drawing was interrupted or shifted so that no region was kept under the microscope light for more than a few minutes.

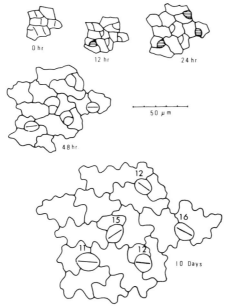

Fig. 10. Six stages in the development of a small region of pea epidermis (part of the region shown in Figs. 9 and 11). Observations of living material using an epi-illumination microscope with an attached drawing apparatus. Observations were started when the stipule was 6mm long, somewhat less than half its final length. All other observations date from this time. Incomplete lines at 0-hr mark the cell walls which were present at the time the tissue was first observed but were clearly the last to form. Striped small cells had a shiny, bulging appearance which according to previous experience is found only in cells which develop into stomata. Note that neighboring cells divided unequally and produced stomata which were separated by at least one epidermal cell.

Fig. 9 shows the mature region whose development will serve as the basis for the discussion of the formation of spacing patterns in the pea epidermis. Stages in the development of a small part of this region, chosen because it appears to represent the elementary processes, are shown in Fig. 10. Stomata numbers 12 and 15 were produced from unequal divisions which occurred in neighboring cells. The orientation or polarity of the developmental events was the same in these neighboring cells, and when the stomata matured the distance between them was about 35μm. This is a distance which approaches the distance found by measurement (Fig. 6) between stomata whose formation is not correlated. It may be concluded that the basic developmental pattern found in peas is different from that of *Sedum* in details but not in principle. An orderly developmental sequence of cell divisions can account for a considerable part of the regularity found in the mature epidermis.

The degree to which this conclusion is general can be judged from

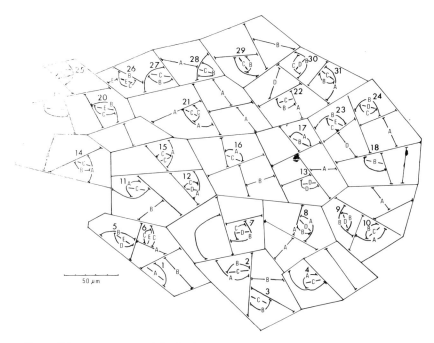

Fig. 11. Schematic summary of the developmental information concerning the region shown in Fig. 9. Cell walls formed during equal divisions were drawn as straight lines and cell walls formed during unequal divisions as curved lines. The endings of the cell walls show the sequence of their formation relative to one another. The letters show the time they were first observed as new cell walls: no letter, at 0-hr, when the epidermis was first observed (see Fig. 10); "A" 12-hr later; "B" at 24-hr; "C" at 48-hr; "D" at 72-hr and "E" after 1 week, when the leaf was essentially mature.

Fig. 11, which is a diagrammatic representation of the developmental events in the entire region studied (shown in Fig. 9). For the sake of simplicity, this figure supplies information concerning the relative timing and orientation of the equal and unequal divisions and excludes all other facts. Two general conclusions will be drawn from this figure:

(a) There is no fixed orientation to the events of stomata development even in a region as small as that shown in Figs. 9 and 11. Though general trends can be found in patches including 4 or 5 stomata, the polarity of the first unequal divisions may even be apposed in neighboring cells, as in stomata 9 and 10. Yet subsequent developmental events were such that no stomata were formed touching one another. The closest stomata in the region studied, numbers 5 and 6, are those which matured late and the cells between them did not grow much before the entire leaf matured.

(b) The second general conclusion is that the processes of stomata formation are extremely variable even when a group of neighboring stomata is considered. Variations of stomata development on the same leaf has already been noticed from studies of isolated stages of development (for example, see Paliwal, 1965, and Inamdar, 1968). The developmental study reported here shows that this variability can be expressed in: (1) The number of unequal divisions which led to stomata formation (generally two divisions, but apparently one for stomata 13, 22 and 26). (2) The location and orientation of the second unequal division relative to the first (compare stomata 20, 24 and 28 with other stomata). (3) The orientation of the final, equal division relative to the orientation of the unequal divisions (compare stomata 24 and 28; 11 and 12). (4) The length of time stomata development required and the period between consecutive divisions (the common case was about a day between stomata, as in numbers 2, 7, 22 and 31, but longer periods were also observed, as in numbers 5 and 8).

It may be concluded that orderly clone formation plays a major role in determining the spacing in *Pisum*, as it does in *Sedum*. Orderly development, however, can not be the only factor involved. The formation of the pattern even though there is no uniform polarity to the developmental events and the great variability in the processes of stomata formation point to the possibility that additional factors could be involved.

V. THE POSSIBILITY OF STOMATA-MESOPHYLL INTERACTIONS

The evidence presented above thus indicates that individual stoma formation is not completely independent of events in the surrounding epidermal and internal tissues. This conclusion is also supported by additional evidence: for example, immature stomata in some mature leaves are found preferentially where stomata density is high (Sachs and Benouaiche, in preparation). More direct evidence was found in *Anagallis,* in which the "stomata-free" region is somewhat larger on the upper side of the leaf even though there are no corresponding differences in cell size and stoma development (Marx and Sachs, 1977). The possible interactions of stomata with the surrounding tissues will therefore be considered below; we will start with the correlation which should be amenable to direct observation.

The most prominent internal factor could be the vascular tissues. Stomata are in fact absent in many plants above the major veins, but these are not common or dense enough to have a major effect on stomata spacing patterns. The relation between the orientation of the stoma pore and the axis of the veins has been considered by Goebell (1922) and Smith (1935). They found that there is a relation between the stomata and the veins but it is not a strict one and does not hold in every case. This variability was considered to reflect the differences in the time of stomata maturation; the correlation might be best only between stomata and veins which develop at the same time; many stomata develop later, when other factors may be active in the same location. In the region studied in *Pisum* the minor veins were marked in Fig. 9. No clear effect of the veins on the occurrence or orientation of the stomata can be seen.

Interactions affecting the location of every stoma could involve the internal parenchyma, the mesophyll, which is always found below the epidermis. This possibility is supported by the presence of a sub-stomatal chamber, or air space, below the epidermis next to each stoma. It has been suggested that the location of this chamber determines the development of the stoma (Pfitzer, 1870; Tomlinson, 1971). However, early stages of stomata development above a mesophyll cell rather than a chamber or an anticlinal cell wall have been observed in various plants (Goebell, 1922; Bunning and Sagromsky, 1948; Jantsch, 1959; Korn, 1972). In these cases it was thought that a sub-stomatal chamber does eventually form, and this chamber must be determined by some influence coming from the stomata, rather than vice versa.

In the case of *Pisum,* it appeared possible that the observed variability of stomata development (Fig. 11) reflects interactions and adjustments leading to an orderly relation between the stomata and the sub-stomatal chamber. This possibility would predict some sort of defined relation between the developing or mature stomata and the cellular configuration in the adjoining mesophyll. In the mature leaf the relation is three dimensional and hard to illustrate, but no such orderly relation could be found in cleared material. There are many air spaces which are not associated with stomata. In the developing epidermis the cells are closer together and the relations are simpler. Fig. 12 is a drawing of a section showing both the developing epidermis and the adjoining mesophyll. No consistent relation can be observed in this case either: future stomata often appear next to the point where three mesophyll cells meet, which might be expected to indicate the location of a future sub-stomatal chamber. This, however, is not the rule and is not even true for the first stomata to mature in the region drawn.

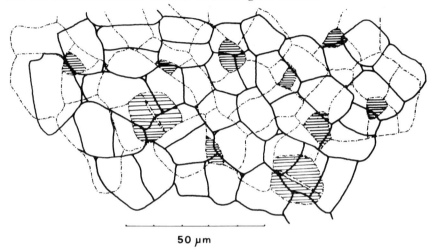

50 μm

Fig. 12. Exact drawing of a section through a developing pea stipule showing the relation between the epidermal cells (broken lines) and the mesophyll cell (entire lines). Stomata and small cells which could be expected to develop to stomata are striped. Note that there is no consistent relation between the developing stomata and the mesophyll.

Usual paraffin procedure was used; section cut at the plane of the surface, 20μm in thickness. Regions were found in which one could see both the young lower epidermis and the developing mesophyll (the future spongy tissue).

It may be concluded that the pattern of mesophyll cells is not a major controlling factor of exact stoma location and of the distances between stomata. However the possibility that interactions between the stomata and the mesophyll cells play some role in controlling both their patterns can not be ruled out.

VI. POSSIBLE EPIDERMAL INTERACTIONS CONTROLLING STOMATA SPACING

Finally, we may consider the possibility that a stoma has a direct effect on the differentiation events in its vicinity. This is the idea of "inhibition", suggested by Bunning and Sagromsky (1948, Bunning, 1948, 1965), which is commonly accepted as obviously correct. The original suggestion (Bunning and Sagromsky, 1948) was that the developing stomata produce a cell division factor, or wound hormone, which inhibits the differentiation of stomata in neighboring cells. The evidence, however, is rather meager and indirect, not only for this specific suggestion but also for the existence of any sort of inhibitory interaction. What might be the best indication for such inhibition is the finding by Bunning and Beigert (1953) that two unequal divisions can occur in the same onion epidermal cell but even then only one stoma is finally formed. The processes leading to this result, however, were not followed but rather reconstructed on the basis of various stages of development found in maturing leaves.

Bunning (1948, 1965) suggested that further evidence for mutual inhibition can be found in leaves in which the stomata do not mature together. He thought that when the leaf is young there is space for only a few stomata and as additional tissue is formed between these early stomata there is uninhibited space in which additional stomata can form. This process may be repeated a number of times. Various relations between the rates of stomata maturation and leaf growth have been recorded by Leick (1955) for different plants, but these neither support nor contradict Bunning's suggestion. If this hypothesis be correct, however, the first stomata to differentiate should form a spacing pattern in the mature leaf, the distances between them being much greater than between later formed stomata. Concrete evidence concerning this prediction can be found in Fig. 11. It can be seen that in *Pisum* there is no obvious order or pattern of stomata maturation, except that groups of 3 or 4 stomata generally matured together (for example, numbers 2, 3 and 4; 7, 8, 12 and 13; 11, 14, 15 and 21). The evidence (Fig. 11) thus points to the maturation of stomata in patches and does not indicate any inhibitory influence of the early stomata or their surroundings. Maturation in patches has been suggested as the basis of the formation of vascular networks (Sachs, 1975) and need not have any direct relation to the formation of spacing patterns of stomata.

A phenomenon which would appear to contradict the idea of inhibitory interactions between stomata is the formation of groups or clusters of stomata in many plants (Weber, 1949), such as the genus *Begonia*. Within these groups, however, the stomata are spaced and do not touch one another. Bunning and Sagromsky (1948) suggested that the formation of groups is associated with the presence of a large sub-stomatal chamber below them and this isolates the stomata and prevents the expression of their inhibitory effects. Sagromsky (1949), however, found group formation in species of *Sedum* in which no pronounced sub-stomatal chamber is present. Weber (1949) suggested that groups form because of an additive effect of the inhibitory influence of their members. Since the stomata of groups often form by repeated divisions of the same mother cell, a special type of clone formation might be involved which would merit further work.

Stomata may influence neighboring cells in ways which are not expressed in the prevention of stomata differentiation and yet control the formation of a spacing pattern. A polarization of differentiation and oriented growth in the cells neighboring on a developing stoma could also influence the distance between stomata in the mature epidermis. An indication that such polarization processes can occur is the finding of Bunning and Sagromsky (1948) that the nuclei of epidermal cells are closer to a neighboring developing stoma than would be expected if their location depended on chance. Another indication of a polarizing effect of developing stomata on the events in neighboring cells is found in various monocotyledons, and especially in grasses (Stebbins and Jain, 1960; Stebbins and Shah, 1960). In these plants the presence of a stomata mother cell is associated with oriented unequal divisions in the neighboring cells which lead to the formation of specialized subsidiary cells next to the stoma. There are no clear reports, however, of polarization effects of a stoma which influence the minimal distance between neighboring stomata.

Evidence for such an effect of developing stomata on growth and differentiation in neighboring cells can be sought in the developmental study of *Pisum* epidermis summarized in Fig. 11. As critical cases which could indicate inhibition of differentiation or polarization of events in neighboring cells we will use the unequal division which did not lead to a mature stoma, the cell marked "A" in Fig. 13, and the development of two small cells which touched one another into stomata numbers 9 and 10. Fig. 13 shows significant stages in the development of the region which included both these events.

When the region shown in Fig. 13 was first observed (0 hr.) a stoma appeared to be forming in the location marked "A" in later stages. In the following days, however, unequal divisions cut off cells right next to "A", and "A" itself became an epidermal cell. Whether a somatic mutation or some other change occurred in "A" can not be determined. The events only point to the possibility of a regulatory process, including interactions, which might occur during stomata development, but do not supply any concrete evidence for an effect of a stoma on neighboring cells.

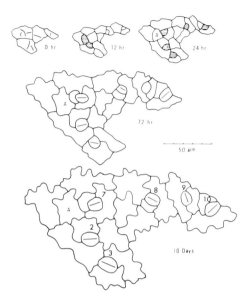

Fig. 13. Stages in the development of part of the epidermis shown in Figs. 9 and 11. All details as in Fig. 10. Note that cell "A" developed from the smaller product of an unequal division and the possibility that stomata numbers 9 and 10 interacted during their development.

Stomata numbers 9 and 10 in Fig. 13 were formed from unequal divisions which did not occur at the same time. It follows that if the earlier division had an effect on events in neighboring cells this was not expressed in the location of the later unequal division, leading to the formation of stoma number 9. The maturation of stoma number 9 also took place in the presence of a mature stoma, number 10, relatively near by. There is some indication of regulatory effects of developing stomata on their surroundings in the orientation of the second unequal divisions which placed the small cells, and thus the future stomata, at maximal distance from one another. The mature stomata were about 30μm apart, which is not an unusual distance judging from the measurements of Fig. 6.

It can be concluded that inhibitory or polarization effects of a stoma on the surrounding cells might occur in *Pisum*. These possible effects of a developing stoma, however, are not very pronounced and a clear proof of their existence is still lacking. Proofs should probably be sought in other plants, in which such control by developing stoma of the development of surrounding cells might be easier to observe.

VII. CONCLUSIONS

The following general picture of the processes which determine the spacing of stomata emerges from this study:

(a) The meristematic cells which form the epidermis go through a stage in which they can take one of two alternative developmental pathways: one leading to the formation of epidermal cells and the other leading to the formation of both epidermal cells and stomata. This stage is not passed by the entire future epidermis at the same time and it often occurs before cell division has ceased.

(b) Whether a cell takes the developmental path leading to stoma formation does not depend on the developmental events in its immediate neighbors. Cells which are in direct contact with one another often follow the same developmental pathway, and the choice of this pathway does not affect the spacing pattern of the stomata.

(c) The development leading to stoma formation also leads to the formation of one or more epidermal cells. The relative position of a stoma and the epidermal cells produced with it depend on the general polarizing effects in the leaf, and are often the same for neighboring cells. This type of orderly development produces stomata which are separated by one or more epidermal cells, the exact number depending on the species. This orderly development is the main basis for the formation of spacing patterns of stomata, regardless of whether the development of the stomata is synchronous or not.

(d) Though the rules of stomata development are the basis of spacing pattern formation, the processes involved are not necessarily determinate. Evidence for the possibility of regulation is found in the variability of stomata development and in the failure of some unequal divisions to lead to stoma formation. It is possible that this expresses some regulatory interactions between the developing stomata and the mesophyll or directly between the developing stomata. These possibilities require further study.

(e) The density of the stomata depends on the number of meristematic cells which take the developmental pathway leading to stoma formation. The density might vary greatly without change in the spacing pattern. While the choice of the exact cells which take this pathway might depend on chance, their number probably depends on the state of the entire region (Marx and Sachs, 1977).

(f) The nature of the internal controls of the processes of stomata formation are completely unknown, as are all cases of this type. It might be significant, however, that the first stage of stoma formation is the polarization of a cell (leading to an unequal division). Cell polarization has also been found to be the determining stage in the formation of organized strands (Sachs, 1969, 1975).

(g) Finally, we might compare the picture of spacing pattern formation arrived at here with the one generally considered in the scientific literature. The difference is not very great, but many workers on pattern formation, such as Turing (1952), Bunning (1965), and Gierer and Meinhardt (1972) assume the existence of a space and ask how this space becomes divided in an orderly way. Stomata patterns suggest that both space and pattern are formed together, and the processes can not be separated. This is probably true for other cases of pattern formation in plants, such as phyllotaxis (Snow, 1955) and the distribution of strands (Sachs, 1975).

ACKNOWLEDGMENTS

I am indebted to D. Cohen and I. Noy-Meir for discussions which were essential for the development of the method used for Figs. 5 and 6 and to R. Gordon for pointing out that these are "radial distribution functions" used in physics. Discussions with A. Fahn concerning stomata development were very valuable. The method of following the development of stomata by gelatin prints was developed during a stay in the laboratory of I. M. Sussex, whose help is gratefully acknowledged. P. Benouaiche skillfully and patiently helped in collecting the data on which Figs. 5 and 6 are based and prepared the sections from which Fig. 13 was drawn.

REFERENCES

Bunning, E. (1948). "Entwicklungs-und Bewegungsphysiologie der Pflanze" Springer-Verlag, Berlin.
Bunning, E. (1965). *Handb. Pflauzenphysiol.* XV (1), 383-408.
Bunning, E. and Biegert, F. (1953). *Z. Bot.* **41**, 17-39.
Bunning, E. and Sagromsky, H. (1948). *Z. Naturforsch.* **3b**, 203-216.
Clark, P. J. and Evans, F. C. (1954). *Ecology* **35**, 445-453.
Claxton, J. H. (1964). *J. Theor. Biol.* **7**, 302-317.
De Bary, A. (1877). "Vergleichende Anatomie der Vegetationsorgane der Phanerogamen und Farne". Verlag W. Engelmann, Leipzig.
Gierer, A. and Meinhardt, H. (1972). *Kybernetik* **12**, 30-39.
Goebell, K. (1922). "Gesetzmassigkeiten im Blattaufbau". Verlag Gustav Fischer, Jena.
Gopal, B. V. and Shah, G. L. (1970). *Am. J. Bot.* **57**, 665-669.
Inamdar, J. A. (1968). *Proc. Ind. Acad. Sci.* **67B**, 157-164.

Jantsch, B. (1959). *Z. Bot.* **47**, 336-372.

Korn, R. W. (1972). *Ann. Bot.* **36**, 325-333.

Korn, R. W. and Fredrick, G. W. (1973). *Ann. Bot.* **37**, 647-656.

Kuhn, A. (1965). "Vorlesungen uber Entwicklungsphysiologie". Springer-Verlag, Berlin.

Lang, A. (1973). *Brookh. Symp. Biol.* **25**, 129-144.

Lawrence, P. A. and Hayward, P. (1971). *J. Cell Sci.* **8**, 513-524.

Leick, E. (1955). *Flora Jena* **142**, 45-64.

Marx, A. and Sachs, T. (1977). *Bot. Gaz.* (in press).

Paliwal, G. S. (1965). *Phytomorph.* **15**, 50-53.

Pfitzer, E. (1870). *Jb. Wiss. Bot.* **7**, 532-560.

Sachs, T. (1969). *Ann. Bot.* **33**, 262-275.

Sachs, T. (1974). *Bot. Gaz.* **135**, 314-318.

Sachs, T. (1975). *Ann. Bot.* **39**, 197-204.

Sagromsky, H. (1949). *Z. Naturforsch.* **4b**, 360-367.

Sinnott, E. W. (1963). "The problem of organic form". Yale U. Press, New Haven.

Smith, G. E. (1935). *Ann. Bot.* **49**, 451-477.

Snow, R. (1955). *Endeavor,* **14**, 190-199.

Stebbins, G. L. and Jain, S. K. (1960). *Develop. Biol.* **2**, 409-426.

Stebbins, G. L. and Shah, S. S. (1960). *Develop. Biol.* **2**, 477-500.

Tomlinson, P. B. (1971). *Adv. Bot. Res.* **3**, 207-292.

Turing, A. M. (1952). *Phil. Trans. Roy. Soc. London,* **B237**, 37-72.

Weber, H. (1949). *Sitzungsber. Heidelberger Akad. Wiss. Math.-natuwiss. KL., Jahrgang 1949,* **6 Abh.,** 161-188.

Wigglesworth, V. B. (1940). *J. Exp. Biol.* **17**, 180-200.

Wolpert, L. 1971. *Curr. Topics Develop. Biol.* **6**, 183-224.

Epigenetic Clonal Variation in the Requirement of Plant Cells for Cytokinins

Frederick Meins, Jr.

*Departments of Botany
and
Genetics and Development
University of Illinois
Urbana, Illinois*

Andrew N. Binns*

*Department of Biology
Princeton University
Princeton, New Jersey*

I. INTRODUCTION

In the development of animals, different regions of the embryo become progressively committed to particular fates. This process, called *determination*, involves stable changes in phenotype that persist in the absence of the factors that initiated the change, and, in some cases, are inherited by individual cells (Cahn and Cahn, 1966; Coon, 1966; Gehring, 1968). An analogous process occurs in the post-embryonic

*Present address: The Rockefeller University, New York, N.Y. 10021. Abbreviation: CDF, cell division factor(s).

development of higher plant species (Heslop-Harrison, 1967; Steeves and Sussex, 1972; Graham and Wareing, 1976). The plant body is derived from shoot and root meristems established during embryogenesis. Shoot and root apices containing the meristems continue to form roots and shoots, respectively, when cultured on the appropriate media (Torrey, 1954; Reinhard, 1954; Ball, 1960; Smith and Murashige, 1970). Thus, the two types of meristem, although still embryonic in character, have become committed to different developmental fates.

These states of determination are stable but by no means fixed. Certain plants exhibit distinct juvenile and adult phases of development in which only the adult phase normally flowers (Goebel, 1905; Brink, 1962). In the case of English Ivy, for example, the juvenile phase grows as the familiar creeping vine with palmate lobed leaves and alternate phyllotaxy. These plants occasionally change to the adult form, an erect bush with entire ovate leaves and spiral phyllotaxy, which is able to flower and fruit. Although this change is usually abrupt, the growing shoot meristem sometimes generates a series of transitional forms before becoming completely adult in its potentialities. When cuttings from shoots exhibiting the juvenile, transitional, or adult phases are grafted, they continue to grow and express the phase of the shoot from which they were derived (Brink, 1962). Tissues from the juvenile and adult forms also express different characters when serially propagated in culture on the same medium (Stoutemyer and Britt, 1965). Juvenile tissues grow rapidly and, like juvenile shoots, root freely whereas adult tissues grow slowly and do not root. It appears, therefore, that the juvenile and adult phases are not only stable and potentially reversible, but can also persist in populations of dividing cells.

Several different cell types in a variety of plant species have been shown to be totipotent, i.e., the cells retain the capacity to develop the complete, fertile plant (Reinert, 1968; Vasil and Hildebrandt, 1965; Steward et al. 1966; Takebe et al., 1971; Duffield et al., 1972; Frearson et al., 1973). These findings and results of nuclear transplantation experiments with amphibia (Gurdon and Uehlinger, 1966; Laskey and Gurdon, 1970; Hennen, 1970) provide compelling evidence that most specialized cells in the same organism are genetically equivalent. The fact that the determined state can persist when cells divide poses, therefore, the fundamental problem of how cells with the same genetic constitution can inherit different characters. Changes in cellular heredity of this type, known as *epigenetic changes* (Nanney, 1958), are well documented in certain unicellular organisms where direct genetic analysis of the phenotypically altered cells is possible (Beale, 1958; Nanney, 1968). In the case of multicellular organisms a less rigorous and indirect approach must be used. Here, epigenetic changes are operationally defined; they are heritable and regularly reversible, occur at high rates, and are limited by the genetic potential of the cell (Nanney, 1958; Meins, 1972).

These properties are summarized in the thought experiment shown in Fig. 1 which also serves to distinguish between epigenetic changes and classical genetic mutations (Meins, 1969). The upper left corner of the diagram shows a tissue that expresses a specific cellular phenotype denoted by open circles. When treated with an inducer, I_1, the cells undergo a stable change to a second phenotype (solid circles) which then persists in the absence of added I_1. These cells retain the new phenotype when cloned indicating that their stability results from a change in cellular heredity and not from interactions among the cells in the tissue. Treatment with a second inducer, I_2, results in the stable reversion of the cells to the original phenotype; thus, the overall process involves heritable changes that are potentially reversible. Finally, the complete organism is regenerated from progeny of cells expressing the two phenotypes to show that the altered cells have remained totipotent.

The subject of this paper is *habituation,* a type of cellular heritable change that commonly occurs when plant tissues are propagated in culture. Using the approach outlined in the thought experiment, we will argue that habituation has an epigenetic basis and, hence, provides an *in*

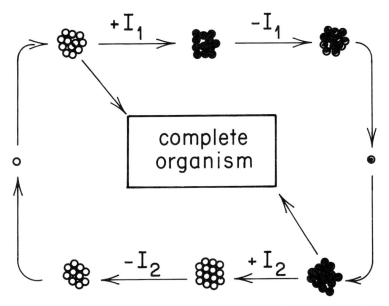

Fig. 1. The properties of epigenetic changes summarized in a thought experiment. Symbols: open and closed circles, cells exhibiting different phenotypes; I_1 and I_2, specific inducers.

vitro model for stable changes encountered in cell determination. We will also propose the working hypothesis that the habituated state is maintained by a positive feedback mechanism.

II. THE NATURE OF THE HERITABLE CHANGE IN HABITUATION

A. *Background*

The proliferation of higher plant cells requires the concerted action of specific growth factors. The most important of these are cell division factors required for cytokinesis and auxins involved primarily in DNA synthesis, mitosis, and cell enlargement (Jablonski and Skoog, 1954; Naylor *et al.*, 1954; Patau and Das, 1961; Simard, 1971). Thus, many plant tissues require cell division factors, usually provided as a synthetic cytokinin, kinetin, and auxin for continuous growth in culture on an otherwise complete medium containing a source of carbon, inorganic salts, and one or more vitamins (Gautheret, 1959). The starting point for our experiments was the observation of Gautheret (1955a) that tissues of wild carrot and *Scorzonera hispanica* in culture gradually lose their requirement for exogenous auxin. Since the heritably altered tissues contain higher concentrations of auxins than the tissue from which they were derived (Kulescha and Gautheret, 1948), apparently the autotrophic cells still require auxin but produce their own supply. This type of heritable change, which Gautheret (1942) called "accoutumance à l'auxine" (habituation), occurs generally. There are reports that cultured tissues from various plant species may habituate for auxin, certain vitamins, and cytokinins (Gautheret, 1955b; Fox, 1963; Street, 1966).

Gautheret (1955a) first proposed that habituation had an epigenetic basis. He argued that it involved "enzymic adaptation" rather than mutation since conversion of tissues to the autotrophic state was gradual and at least partially reversible. Although this view is supported by recent studies showing that entire plants can be regenerated from habituated tissue of single cell origin (Lutz, 1971; Sacristan and Melchers, 1977), these findings and the earlier studies of Gautheret with uncloned tissues do not rule out a mechanism involving genetic mutation followed by cell selection.

B. *Epigenetic changes in cytokinin-habituation*

We attempted to characterize the heritable change in cytokinin-habituation. We chose, as an experimental object, pith parenchyma tissues of Havana 425 tobacco because these tissues habituate rapidly, are readily cloned, and remain organogenic after many transfer generations in culture. Our earlier experiments confirmed by cloning that cytokinin-autotrophy is inherited by individual cells and is an extremely stable phenotype (Binns and Meins, 1973). Epigenetic changes are directed and occur at high rates relative to classical mutation. At the tissue level, cytokinin-habituation is also a rapid process; explants of pith tissue often habituate after a single transfer on a complete medium. Estimating the critical parameter, the rate at which cells habituate, depends upon reliable measurements of the frequency of habituated cells in tissues at different times. A direct approach in which cell suspensions are prepared from the tissue and clones assayed for habituation was not feasible because cloning procedures may select for a particular cell type and are too slow to detect events occurring at the rate anticipated for epigenetic changes. We have recently turned, therefore, to an indirect approach based on the method of Luria and Delbrück (1943) for measuring mutation rates in bacterial populations. Pith explants of Havana 425 tobacco habituate when incubated on medium without added cytokinin at 35°C, about 10°C above the standard culture temperature (Meins, 1974). Under these conditions the cells enlarge considerably and the tissues become translucent. Some of the explants develop dense, white, hyperplastic nodules which are clearly discernible by trans-illumination. These nodules are habituated and grow when subcultured at 25°C without added cytokinin whereas the surrounding translucent tissues do not. Thus, the nodules provide a convenient marker for the habituation event. We calculated rates of habituation from the proportion of cultures without discernible nodules and the number of cell generations during the incubation period. Depending upon the season, pith tissues occasionally habituate even at 25°C. Thus, estimates of rate were corrected for spontaneous conversions occurring in a second set of cultures incubated at 25°C. Results from two experiments are summarized in Table I. The estimates of rate varied widely with an average of >2.8 × 10⁻³ conversions *per* cell generation for five experiments. The actual rate is undoubtedly much higher since habituated cells arising late in the incubation period probably form nodules too small to detect by eye. The important point is that habituation occurs 100 to 1000 times faster than the rates reported for somatic mutation in tobacco, 5-50 × 10⁻⁶ (Sand *et al.*, 1960; Carlson, 1974).

TABLE I

Conversion rate of pith cells to the habituated phenotype[a]

		Experiment	
		12/22/75	5/14/76
Fraction of 25 explants without nodules after 21 days (P_0):	25°C	1.00	0.48
	35°C	0.64	<0.04
Average number of conversions per tissue explant (M):[b]	25°C	0	0.73
	35°C	0.45	>3.26
Cells per explant:	Initial	5,180±430(5)[c]	4,780±169(5)
	25°C	—	6,291± 84(5)
	35°C	5,980±230(4)	5,312±482(5)
Cell generations (G):[d]	25°C	—	2,184
	35°C	1,154	772
Conversion rate per cell generation (=M/G):	35°C (induced)	3.94×10⁻⁴	>4.2×10⁻³
	25°C (control)	0	3.4×10⁻⁴
	Net	3.9 ×10⁻⁴	>3.9×10⁻³

[a]Data from Meins (1977)

[b]Calculated from fraction of explants without nodules assuming a Poisson distribution of conversion events, i.e., $M = -1nP_0$.

[c]Cell number ± standard error for (N) replicate explants without nodules.

[d]Average number of cell generations over 21 days at the temperature indicated, i.e., G = (Final Cell No. — initial Cell No.)/1n 2.

Skoog and Miller (1957) showed that cultured tobacco tissues form organized buds and roots when grown on a medium containing the right combination of cytokinin and auxin. With this approach we obtained 62 complete, fertile tobacco plants from 13 different clones of habituated tissue (Binns and Meins, 1973). The plants regenerated from non-habituated as well as habituated clones had thicker stems, larger leaves, and a darker pigmentation than seed-grown plants raised under comparable conditions. During prolonged culture tobacco cells commonly become polyploid or aneuploid (Murashige and Nakano, 1966; Sacristán, 1967) and it is these changes in chromosomal constitution that lead to morphological aberrations of the type we observed (Zagorska *et al.*, 1974). Although the cultures may have undergone genetic modification, the important point is that cells derived from habituated clones are totipotent. These cells have also lost their habituated character. When pith tissues from the regenerated plants are returned to culture they grow profusely on kinetin-containing medium but not on basal medium. At some point in the regeneration of plants, therefore, cells from habituated clones have reverted to the non-habituated state.

In the experiments described above, plants were regenerated from autotrophic clones subcultured for several passages before induction of organogenesis. It may be argued, therefore, that during prolonged culture nonhabituated cells arise at very low rates by back mutation and that it is these cells that eventually form plants. There are several lines of evidence against this. First, we have subcloned habituated lines and immediately placed the subclones on an inductive medium. In a typical experiment, 33 of 45 subclones formed plantlets. Thus, the plantlets do not develop from a small number of revertant cells arising during prolonged culture. Second, with recently isolated subclones, the rate of bud formation, the first step in regenerating plants, is roughly the same for habituated and non-habituated clones. Finally, we have measured the rate of bud initiation by calculating the number of cell generations calluses undergo before forming their first bud. Subclones from the habituated clone 259H gave an average rate of $8.2 \pm 5.8 \times 10^{-3}$ initiations *per* cell generation ($\pm$ standard error, 7 subclones). This value is at least 1000 times faster than would be expected were bud initiation to depend upon back mutation. It appears, therefore, that the plants either develop directly from habituated cells or from non-habituated cells that arise in the tissues at very high rates.

To summarize, the experiments presented here provide strong evidence that cytokinin-habituation has an epigenetic basis. It occurs at high rates, is regularly reversible, and leaves the heritably altered cell totipotent. Moreover, since cytokinin-habituated cells produce cell division factors (Fox, 1963; Wood *et al.*, 1969; Dyson and Hall, 1972; Einset and Skoog, 1973) of the type produced by certain tissues of the plant (Torrey, 1976), it appears that habituation involves the stable expression of normally silent genetic potentialities of the pith cell. Habituation offers, therefore, an unusual opportunity to study epigenetic changes in a specific differentiated function under precisely controlled conditions in culture. The remaining sections deal with experiments, some still in progress, aimed at characterizing the habituation process at the cellular level and learning how the habituated state is maintained in a population of dividing cells.

III. HABITUATION IS A GRADUAL PROCESS

In the arguments presented so far, we made the simplifying assumption that habituation involves shifts in phenotype between just two states and, in this regard, resembles an "on-off" switch. It turns out, however, that a rheostat is probably a more apt metaphor for the habituation process. Several lines of evidence indicate that tobacco cells in culture exhibit different degrees of cytokinin-autotrophy and, moreover, that the conversion of these cells to the fully autotrophic state is a gradual process (Meins, 1974; Meins and Binns, 1977). Different habituated clones isolated from the same line of tissue differ in their growth rate on basal medium and dose response to the synthetic cytokinin, kinetin (Tandeau de Marsac and Jouanneau, 1972; Binns and Meins, 1973). Fig. 2 shows the results of an experiment in which we surveyed a large number of clones for degree of habituation expressed as the ratio, R, of growth rate without kinetin to growth rate with kinetin added at a concentration optimal for the growth of non-habituated tissues. Although these clones were recently isolated from the same tissue they varied widely in R value suggesting that individual cells differ in degree of habituation.

We performed two types of experiments to find out whether cells are fixed in their state of habituation or vary with time in culture. In short term experiments, subclones with high and low R values were isolated from a cloned line, 156H. The subclones were cloned as soon as possible and the second generation of clones, i.e., subclones of subclones, assayed for habituation. The distributions of R values obtained from highly as well as slightly habituated subclones overlapped extensively but were biased toward the parent clone from which they were isolated (Fig. 3). Thus, 156H-23 with R = 1.05 and 156H-3 with R = 2.75 gave distributions of daughter clones with mean R values of 1.30 and 2.44

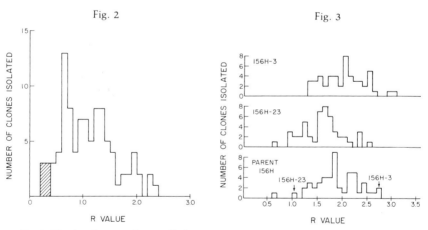

Fig. 2. Clonal variation in degree of habituation. Habituation expressed as R value, the ratio of tissue doubling constants—kinetin/+kinetin; cross-hatched region, non-habituated clones. From Meins (1974).

Fig. 3. Evidence for shifts in degree of habituation: a short term subcloning experiment with 156H. From Meins and Binns (1977).

respectively. Several important conclusions may be drawn from these results: First, since the subclones do not cluster around the R value of the parent clone, apparently cells shift in degree of habituation at a rapid rate relative to the time required for cloning and assaying tissues; second, clones shift to higher as well as lower R values suggesting that transitions among different states of habituation are reversible; and finally, over short periods of time cloned tissues appear to be in histogenetic equilibrium with transitions of cells to higher and lower states of habituation roughly balanced.

In the second type of experiment, we monitored the R value of a cloned tissue serially subcultured on basal medium for long periods of time. A slightly habituated clone, 17H, was used since more highly habituated cells arising in this tissue would have a selective advantage and thus could be detected by increases in R value. This clone gradually increased in R value over a period of two years (Fig. 4A). To find out whether or not the increase reflected gradual changes at the cell level, we cloned 17H at three different times, indicated by arrows in Fig. 4A, and assayed the subclones for habituation. The mean R values of the subclones increased with time and were roughly equal to the R values obtained with the parent tissue at the same time (Fig. 4B). Thus, the

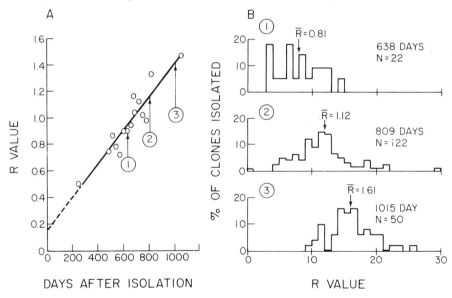

Fig. 4. Gradual and progressive changes in the habituation of clone 17H. A. Changes in the R value of tissues during prolonged culture. Arrows indicate when subclones were isolated. B. The R value distributions of subclones isolated at the time indicated. $\bar{R}$., average R value of the subclones; N, number of subclones tested. From Meins and Binns (1977).

cloning procedure did not select for a particular class of cells or induce detectable changes in degree of habituation. The distribution of subclones apparently reflects the population of cells in the original tissue. The critical observation was that the entire distribution of subclones shifted upward in R value over long periods of time indicating that habituation is gradual and progressive at the cell level. These changes also appear to have an epigenetic basis. Cells exhibiting different degrees of habituation remain totipotent and can still return to the non-habituated state (Meins and Binns, 1977).

IV. A POSITIVE FEEDBACK HYPOTHESIS

The fundamental question that arises is: How are different states of habituation maintained in cells with apparently the same genetic constitution? Stability of this type can be accounted for by the intrinsic properties of interdependent regulatory pathways in the cell (Delbrück, 1949; Monod and Jacob, 1961). The basic idea is that certain systems of chemical reactions are bistable, i.e., they can exist in alternative, stable, steady-states (Kacser, 1963; Rosen, 1972). These states most commonly arise in positive feedback loops in which a reaction intermediate either triggers its own synthesis or inhibits its own degradation. Two genetically equivalent cells with loops of this type could, depending upon their history, inherit different phenotypes. The stability and variety of these phenotypes would depend upon the kinetic properties of the system and hence, ultimately, upon the genetic potentialities of the cell. Although bistability provides an attractive mechanism for epigenetic changes, direct experimental evidence for inheritance of this type has been obtained only in prokaryotes, for example, the "pseudomutation" of the lac operon in galactoside-induced *Escherichia coli* (Novick and Weiner, 1957).

Certain regulatory molecules induce their own production by plant cells, e.g., ethylene by flower tissue of morning glory (Kende and Baumgartner, 1974), cyclic-AMP by aggregating cells of the cellular slime mold, *Dictyostelium discoideum* (Shaffer, 1975), and the auxin, indoleacetic acid, by cultured explants of pith from *Nicotiana glauca* × *langsdorffii* hybrid (Cheng, 1972). We propose that, in a similar way, CDF induces its own production and it is this positive feedback relationship that keeps cells habituated when they divide. Fig. 5 shows an heuristic model for habituation that incorporates this working hypothesis. Assuming that CDF can enter and leave cells and is degraded, the system can exhibit two steady states: one in which the concentration of CDF is zero, the other in which CDF is above some critical threshold

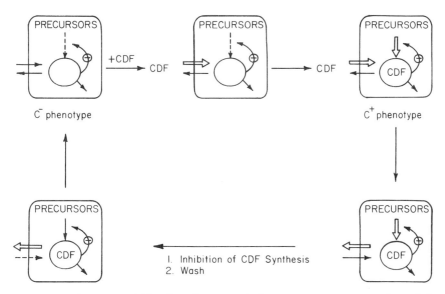

Fig. 5. An heuristic model based on the positive feedback hypothesis. Squares with rounded corners represent cells permeable to CDF (double arrows crossing the cell boundary) with an autocatalytic pathway for CDF production. The arrow with a "+" sign indicates CDF is a positive effector. The CDF pool is represented by a circle. C⁻, non-habituated; C⁺, habituated.

concentration at which cells are triggered to produce CDF. Thus, according to the model, when non-habituated cells (the upper left corner of Fig. 5) are treated with exogenous CDF, they take up the growth factor which triggers CDF production. Thereafter, the cells produce CDF without an exogenous supply of this factor, and, hence, exhibit the habituated phenotype. If the cells are treated to block CDF production and washed to lower the internal CDF concentration below the critical threshold, then the cells complete the cycle by returning to the non-habituated state.

The positive feedback hypothesis makes a specific prediction which we have begun to verify experimentally, *viz.*, it should be possible to shift tissues between the habituated and non-habituated state by momentarily altering the cellular concentrations of CDF. Our strategy was to grow cloned, habituated tissues for several transfers under non-permissive conditions that are thought to prevent CDF production by the cells. The tissues are then shifted to permissive conditions and assayed for habituation. Since there is carryover of growth factors when tissues are subcultured, we ran parallel experiments in which tissues under non-permissive conditions were incubated with optimal concentrations of kinetin and low concentrations of kinetin that support growth but are thought to be below the threshold needed to induce CDF production. Thus, if the hypothesis is correct, tissues on high-kinetin medium should remain habituated when assayed under permissive conditions whereas tissues on low-kinetin medium should shift to the non-habituated state. Moreover, tissues in this state, when transferred on high-kinetin media under permissive conditions, should regain their habituated character.

We attempted to block CDF production in two ways: by cold treatment and by subculturing tissues on media containing low concentrations of auxin. Syōno and Furuya (1971) found that cytokinin-habituated tissues of tobacco reversibly lose their autotrophic phenotype when cultured at 16°C, about 10°C below the standard culture temperature. We confirmed this observation using cloned tissues and showed that habituated cells, which require cytokinins at 16°C, exhibit the same dose response to kinetin as stable, nonhabituated tissues at either 16°C or 25°C. Thus, it appears that cold sensitivity results from a decreased production of CDF rather than from a decreased sensitivity of cellular receptors for this factor. We found that tissues subcultured at 16°C at low kinetin concentrations usually become cold resistant and, hence, could not be used in reversal experiments. Fortunately, one cloned line, 131H, was stable under these conditions. This tissue was subcultured for 14 transfers at 16°C and degree of habituation, expressed as R value, assayed under permissive conditions after each transfer (Fig. 6). Tissues on low-kinetin medium rapidly decreased in R value and were no longer habituated after the seventh transfer. Tissues on high-kinetin medium decreased in R value until the ninth transfer and thereafter increased. Table II summarizes results obtained when tissues, after 10 transfers at 16°C, were subcultured twice on combinations of media with and without kinetin. Tissues grown previously on high-kinetin medium remained habituated although they still increased in degree of habituation when treated with kinetin at 25°C. On the other hand, tissues grown on low-kinetin medium were no longer habituated but could be induced to habituate by a single transfer at 25°C on kinetin medium. Thus, 16°C treatment, presumably by blocking CDF production, reverses habituation, while brief treatment with CDF re-establishes the habituated state.

Cytokinin-habituated tissues of tobacco require an exogenous source of CDF when cultured on media in which the concentration of auxin has been reduced (Tandeau de Marsac and Jouanneau, 1972; Einset, 1977). Attempts to reverse habituation by low-auxin treatment gave erratic results. Significant reversal occurred in only one of the three

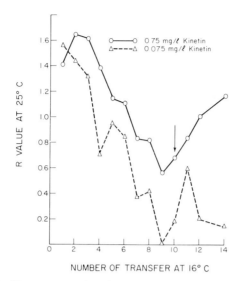

Fig. 6. Effect of cold treatment on the subsequent degree of habituation of clone 131H at 25°C. R values less than 0.4-0.5 are in the range expected for non-habituated tissues (Meins and Binns, 1977). Arrow indicates source of tissues for the re-induction experiments shown in Table II.

TABLE II

Cold treatment of clone 131H:
Reversal and re-induction of habituation.

Treatment at 16°C[a]	Subsequent treatment at 25°C			
	1st transfer		2nd transfer	
	Medium[b]	Growth[c]	Medium	Growth
Low Kinetin	−Kinetin	0.93±0.22†	−Kinetin	0.54±0.04†
			+Kinetin	0.42±0.08†
	+Kinetin	30.66±8.54	−Kinetin	5.39±1.80
			+Kinetin	5.33±0.31
High Kinetin	−Kinetin	6.95±0.68	−Kinetin	6.66±1.04
			+Kinetin	3.91±0.35
	+Kinetin	23.28±6.09	−Kinetin	10.53±1.11
			+Kinetin	5.16±0.22

[a] Tissues after 10 transfers on low-kinetin (0.075 mg/1) or high-kinetin (0.75 mg/1) media.
[b] Tissues grown on basal medium (Binns and Meins, 1973) or basal medium with 0.3 mg/1 kinetin.
[c] Growth expressed as fresh weight after 21 days (W) minus initial fresh weight (W_0) divided by W_0, i.e., $(W-W_0)/W_0$. Mean values ± standard error for 8 replicates.
†Tissue is dead.

experiments performed so far (Table III). Nevertheless, when reversal did occur, the tissues returned to the habituated state after a brief kinetin treatment in agreement with results of the 16°C experiments.

Our experiments provide indirect evidence that the habituated state is maintained by a positive feedback loop involving CDF. We are currently trying to test this hypothesis directly by measuring the CDF content of tissues under conditions that either reverse or induce habituation. The hypothesis predicts that CDF concentrations should decrease in tissues during reversal and that tissues induced to habituate by CDF treatment should produce more CDF than was supplied.

TABLE III

Low-auxin treatment of clone 246H;

Reversal and re-induction of habituation.

Treatment with low auxin[a]	Subsequent treatment on standard auxin medium[b]			
	1st transfer		2nd transfer	
	medium	growth[c]	medium	growth
low kinetin	−kinetin	3.18±0.75	−kinetin	†
			+kinetin	†
	+kinetin	45.19±8.04	−kinetin	30.9±5.4
			+kinetin	21.7±2.8
high kinetin	−kinetin	202.73±20.79	−kinetin	—
			+kinetin	—
	+kinetin	83.53±23.22	−kinetin	—
			+kinetin	—

[a]0.02 mg/1 α-naphthalene acetic acid (NAA), and 0.03 or 0.3 mg/1 kinetin, as indicated.
[b]2.0 mg/1 NAA with and without 0.3 mg/1 kinetin.
[c]Growth expressed as in Table II. Mean value ± standard error for 8 replicates.
†Tissue is dead.

V. CONCLUSIONS

We have obtained strong experimental evidence for the hypothesis advanced by Gautheret (1955a) over 20 years ago that habituation is a progressive, gradual process involving epigenetic changes rather than classical genetic mutations. Our studies bear directly on the problem of how epigenetic changes are initiated and maintained. These changes are defined operationally and do not imply a specific cellular or molecular mechanism. For example, the directed, reversible nature of habituation does not necessarily exclude a conservative genetic mechanism involving directed alterations in the sequential order of genes in the chromosome. There is good genetic evidence from studies of corn that transposable elements, presumably regulatory genes, can migrate to new locations on the chromosome, insert into the DNA, and control the transcription of nearby structural genes (McClintock, 1967; Fincham and Sastry, 1974; Nevers and Saedler, 1977). Phase variation in *Salmonella* appears to be a similar phenomenon (Lederberg and Iino, 1956; Iino, 1969). Here, alternative heritable states result from the inversion of DNA base sequences adjacent to one of the genes controlling the expression of flagellar antigens (Zieg *et al.,* 1977). Genetic rearrangement of this type provides an attractive explanation for epigenetic changes; however, there is, at present, no evidence that inversions and transpositions can be induced and reversed in a directed fashion by specific inducers.

We have proposed what amounts to a "systems" hypothesis for habituation that does not exclude a directed genetic mechanism. Although, for purposes of illustration, our heuristic model specified positive-feedback regulation of CDF production, models with similar properties could have been constructed in which CDF transport or degradation are regulated. The salient feature of our working hypothesis is that CDF regulates its concentration in the cell by some sort of positive-feedback interaction. This hypothesis can account for the stability of the habituated state and is supported by experimental evidence. It does not, however, provide an adequate explanation for different degrees of habituation, the progressive nature of the cellular changes, or the observation that certain non-habituated cell lines, although maintained on a kinetin-containing medium, habituate very slowly. Nevertheless, the hypothesis has led to experimental procedures for shifting cells at will between the cytokinin-dependent and habituated states under precisely defined conditions and provides a framework for asking more detailed questions regarding molecular events underlying the habituation process.

ACKNOWLEDGMENTS

Original work reported here was supported by grants CA 13587 and CA 20053 from the National Cancer Institute, U.S. Public Health Service, BMS 74-18906 from the National Science Foundation, and the Anita G. Mestres Cancer Fund, Princeton University.

REFERENCES

Ball, E. (1960). *Growth* **24**, 91-110.

Beale, G. H. (1958). *Proc. Roy. Soc. London B* **148**, 308-314.

Binns, A. and Meins, F., Jr. (1973). *Proc. Nat. Acad. Sci. U.S.* **70**, 2660-2662.

Cahn, R. D., and Cahn, M. B. (1966). *Proc. Nat. Acad. Sci. U.S.* **55**, 106-114.

Carlson, P. S. *(1974). Genetical Res. Cambridge* **24**, 109-112.

Cheng, T. Y. (1972). *Plant Physiol.* **50**, 723-727.

Coon, H. G. (1966). *Proc. Nat. Acad. Sci. U.S.* **55**, 66-73.

Delbruck, M. (1949). Discussion following the paper by Sonneborn and Beale. *Colloq. Internat. Centre Nat. Recherche Sci.* **7**, 25.

Duffield, E. C. S., Waaland, S. D., and Cleland, R. (1972). *Planta* **105**, 185-195.

Dyson, W. H. and Hall, R. H. (1972). *Plant Physiol.* **50**, 616-621.

Einset, J. W. (1977). *Plant Physiol.* **59**, 45-47.

Einset, J. W. and Skoog, F. (1973). *Proc. Nat. Acad. Sci. U.S* **70**, 658-660.

Fincham, J. R. S. and Sastry, G. R. K. (1974). *Ann. Rev. Genet.* **8**, 15-50.

Fox, J. E. (1963). *Physiol. Plantarum* **16**, 793-803.

Frearson, E. M., Power, J. B., and Cocking, E. C. (1973). *Develop. Biol.* **33**, 130-137.

Gautheret, R. J. (1942). *Bull. Soc. Chim. Biol. Paris* **24**, 13-47.

Gautheret, R. J. (1955a). *Rev. Gén. Bot.* **62**, 1-110.

Gautheret, R. J. (1955b). *Annu. Rev. Plant Physiol.* **6**, 433-484.

Gautheret, R. J. (1959). "La Culture des Tissus Végétaux." Masson, Paris.

Gehring, W. (1968). *In* "The Stability of the Differentiated State" (H. Ursprung, ed.), pp. 134-154. Springer Verlag, Berlin.

Goebel, K. (1905). "Organography of Plants." (I. B. Balfour, trans.), vol. 1. The Clarendon Press, Oxford.

Graham, C. F. and Wareing, P. F. (1976). *In* "The Developmental Biology of Plants and Animals." (C. F. Graham and P. F. Wareing, eds.), pp. 45-54. W. B. Saunders, Philadelphia.

Gurdon, J. B. and Uehlinger, V. (1966). *Nature* **210**, 1240-1241.

Hennen, S. (1970). *Proc. Nat. Acad. Sci. U.S.* **66**, 630-637.

Heslop-Harrison, J. (1967). *Ann. Rev. Plant Physiol.* **18**, 325-348.

Iino, T. (1969). *Bact. Rev.* **33**, 454-475.

Jablonski, J. R. and Skoog, F. (1954). *Physiol. Plantarum* **7**, 16-24.

Kacser, H. (1963). *In:*"Biological Organization at the Cellular and Supercellular Level." (H. Harris, ed.), pp. 25-41. Academic Press, New York.

Kende, H. and Baumgartner, B. (1974). *Planta* **116**, 279-289.

Kulescha, Z. and Gautheret, R. (1948). *C. R. Acad. Sci. Paris* **227**, 292-294.

Laskey, R. A. and Gurdon, J. B. (1970). *Nature* **228**, 1332-1334.

Lederberg, J. and Iino, T. (1956). *Genetics* **41**, 743-757.

Luria, S. E. and Delbrück, M. (1943). *Genetics* **28**, 491-511.

Lutz, A. (1971). *In* "Les Cultures de Tissus de Plantes; *Colloq. Internat. Centre Nat. Recherche Sci., Paris*", No. **193**, pp. 163-168. Centre National de la Recherche Scientifique, Paris

McClintock, B. (1967). *Develop. Biol., Suppl.* **1**, 84-112.

Meins, F., Jr. (1969). "Control of Phenotypic Expression in Tobacco Teratoma Cells." Ph.D. Diss., The Rockefeller University, New York.

Meins, F., Jr., (1972). *Prog. Exp. Tumor Res.* **15**, 93-109.

Meins, F., Jr., (1974). *In:*"Tissue Culture and Plant Science 1974" (H. E. Street, ed.), pp. 233-264. Academic Press, New York.

Meins, F., Jr. (1977). *Am. J. Pathol.* **89**, 687-702.

Meins, F., Jr. and Binns, A.N. (1977). *Proc. Nat. Acad. Sci. U.S.* (1977). **74**, 2928-2932.

Monod, J. and Jacob, F. (1961). *Cold Spr. Harb. Symp. Quant. Biol.* **26**, 389-401.

Murashige, T. and Nakano, R. (1966). *J. Hered.* **57**, 115-118.

Nanney, D. L. (1958). *Proc. Nat. Acad. Sci. U.S.* **44**, 712-717.

Nanney, D. L. (1968). *Science* **160**, 496-502.

Naylor, J., Sander, G. and Skoog, F. (1954). *Physiol. Plantarum* **7**, 25-29.

Nevers, P. and Saedler, H. (1977). *Nature* **268**, 109-115.

Novick, A. and Weiner, M. (1957). *Proc. Nat. Acad. Sci. U.S.* **43**, 553-566.

Patau, K., and Das, N. K. (1961). *Chromosoma* **11**, 553-572.

Reinert, J. (1968). *Naturwissenschaften* **4**, 170-175.

Reinhard, E. (1954). *Z. Bot.* **42**, 353-376.

Rosen, R. (1972). *In:*"Foundations of Mathematical Biology." Vol. 2, pp. 79-140. Academic Press, New York.

Sácristan, M.D. (1976). *Molec. Gen. Genet.* **99**, 311-321.

Sácristan, M. D. and Melchers, G. (1977). *Molec. Gen. Genet.* **152**, 111-117.

Sand, S. A., Sparrow, A. H. and Smith, H. H. (1960). *Genetics* **45**, 289-308.

Shaffer, B. M. (1975). *Nature* **255**, 549-552.

Simard, A. (1971). *Can J. Bot.* **49**, 1541-1549.

Skoog, F. and Miller, C. O. (1957). *Soc. Exp. Biol. Symp.* **11**, 118-131.

Smith, R. H. and Murashige, T. (1970). *Am. J. Bot.* **57**, 562-568.

Steeves, T. A. and Sussex, I. M. (1972). "Patterns in Plant Development" Prentice-Hall, Englewood Cliffs.

Steward, F. C., Kent, A. E., and Mapes, M. O. (1966). *Current Topics Develop. Biol.* **1**, 113-154.

Stoutemyer, V. T. and Britt, O. K. (1965). *Am. J. Bot.* **52**, 805-810.

Street, H. E. (1966). *In:*"Cells and Tissues in Culture" (E. N. Willmer, ed.), vol. 3, pp. 533-629. Academic Press, London.

Syōno, K. and Furuya, T. (1971). *Plant Cell Physiol.* **12**, 61-71.

Takebe, I., Labib, G. and Melchers, G. (1971). *Naturwissenschaften* **58**, 318-320.

Tandeau de Marsac, N. and Jouanneau, J. P. (1972). *Physiol Veg.* **10**, 369-380.

Torrey, J. G. (1954). *Plant Physiol.* **29**, 279-287.

Torrey, J. G. (1976). *Ann. Rev. Plant Physiol.* **27**, 435-459.

Vasil, V. and Hildebrandt, A. C. (1965). *Science* **150**, 889-892.

Wood, H. N., Braun, A. C., Brandes, H., and Kende, H. (1969). *Proc. Nat. Acad. Sci. U.S.* **62**, 349-356.

Zagorska, N. A., Shamina, Z. B., and Butenko, R. G. (1974). *Biologica Plantarium* **16**, 262-274.

Zieg, J., Silverman, M., Hilmen, M., and Simon, M. (1977). *Science* **196**, 170-172.

IV. Nuclear and Genetic Events
in Clone Initiation

Clonal Analysis of Development: X-Inactivation and Cell Communication as Determinants of Female Phenotype

Barbara R. Migeon

Department of Pediatrics
Johns Hopkins University School of Medicine
Baltimore, Maryland 21205

I. INTRODUCTION

There is abundant evidence that only a single X is functional in human somatic cells (see review by M. Lyon, 1972). As a consequence of an event in early embryonic development, females are mosaic with regard to activity of their X-linked genes; some cells express the maternal allele, others the paternal one. It is not clear whether the primary event is inactivation of a single X in each cell, as suggested by Lyon (1961), Russell (1961), and others, or if only a single X is activated in somatic cells, all others becoming inert as a consequence of having not been turned on. In any event, the resultant mosaicism has provided a means to analyze the events in development, and studies of cellular mosaicism have been applied to a series of pertinent problems.

Human cells as an experimental model. The studies that I will describe involve experiments using human somatic cells in culture as probes for events occurring during development. Naturally occurring mutations, many unique to man, have resulted in variations in enzyme phenotype demonstrable at the level of single somatic cells. These variants are stable in cell cultures, established from biopsies of tissues which express the mutation. Human cells in culture are unique among mammalian cells because they do not spontaneously transform *in vitro* and do maintain precisely normal karyotypes. Their 24-hour generation time provides abundant progeny for study. Moreover, these cells can be subjected to experimental manipulation. Colonies consisting of progeny from single cells can be obtained from dilute suspensions of non-aggregating cells by plating aliquots of 10 cells into plastic petri dishes and incubating for approximately 10 days. Well-isolated colonies containing 1000-2000 cells can be picked using cloning cylinders (Ham and Puck, 1962) and new cultures established consisting of a population of cells all derived from one original cell.

G6PD as probe extraordinaire. Glucose-6-phosphate dehydrogenase (G6PD), an X-linked enzyme in all mammals, has been used as an effective probe for the clonal analysis of events in development. The frequent occurrence of two forms of the enzyme (G6PDA and G6PDB), differing in electrophoretic mobility, provides a means of determining the number and parental origin of the active X chromosome(s) in any cell. As a consequence of X inactivation, one X chromosome becomes the sole determinant of the X-specified characteristics of the cell. Clonal populations of skin cells, *heterozygous* for the variants, express only a single form of the enzyme (Davidson et. al., 1963). This indicates not only a single active X in each somatic cell, but also that the inactivation is fixed so that all progeny of a single cell have the same active X. When there are two active X chromosomes in the same cell (i.e. as a result of cell fusion) and each X carries a different variety of G6PD, the hybrid cells express not only both parental isozymes, but also a hybrid molecule, the heteropolymeric form of the enzyme made up of association of the parental subunits. The presence of the heteropolymer is a sensitive indicator of two functional X chromosomes in a single cell and, therefore, can be used as a probe for cells with two active X chromosomes.

The cellular mosaicism in heterozygotes carrying G6PD variants has been used to investigate the origin of various human tumors. Analysis of tumors from heterozygous females has revealed a monoclonal origin for some and a multicellular origin for others, providing important clues to the nature of cells from which the neoplasms arise (Linder and Gartler, 1965; Fialkow, 1976).

Another useful locus, hypoxanthine guanine phosphoribosyl transferase (HGPRT). The HGPRT locus has provided an additional sensitive probe. In this case, there is no heteropolymer since the mutation in man leads to enzyme deficiency (Seegmiller et. al., 1967); but the enzyme phenotype *is* demonstrable at the level of single cells. In addition, nutrient media are available which will disfavor or favor the growth of wild-type cells over those of mutant type. HGPRT deficient cells can be eliminated in the presence of HAT medium, but proliferate in medium containing purine analogs toxic to wild-type cells (Szybalski et. al., 1962).

Molecular basis for single active X. The molecular basis for the single active X remains to be elucidated. Studies utilizing clonal probes have revealed some insights. Analysis of G6PD variants in interspecific hybrids (mouse x human) or human intraspecific hybrids (G6PDA x G6PDB) indicates that the two X chromosomes can function in the same somatic cell and that one X does not turn off the activity of another (Migeon, 1972; Migeon et. al., 1974), nor activate a silent X.[*] Treatment of clones from heterozygotes which express only a single G6PD variant have indicated that the inactivation phenotype is a stable one; neither hormones, polyanions, Simian Virus 40, dimethylsulfoxide, nor bromodeoxyuridine alter the phenotype of the clone under treatment (Comings, 1966; Migeon, 1972; Romeo and Migeon, 1975; and Migeon, unpublished observations). Selective pressure for an allele on the inactive X chromosome is ineffective, even when lack of activation leads to cell death (Migeon, 1972).

Extent of inactivation. Analyses of clonal populations from heterozygotes for X-linked variants have been used to determine whether a whole X chromosome is inactivated or only parts thereof. There are more than 100 loci on the human X chromosome (McKusick, 1975), but only a handful of these involve mutations demonstrable at the level of single cells. Of the seven which are amenable to clonal analysis, all have shown two populations of clones with respect to the variant — one expressing the normal allele and the other expressing the variant — evidence for only one active allele at each locus (see Migeon, 1977 — Table I). The loci analyzed have either been assigned to the long arm of

[*]The possibility that hybridization may result in occasional localized derepression (Kahan and DeMars, 1975) needs further exploration.

the human X chromosome or are yet unassigned; therefore, it is not clear whether loci on the short arm of the X are subject to inactivation. The possibility exists that some X loci escape inactivation, especially because individuals with X-chromosomal aneuploidy have somatic as well as gonadal abnormalities. Individuals with multiple X chromosomes often manifest mental retardation and skeletal anomalies which are more severe as the number of X chromosomes per cell increases (Ferguson-Smith, 1965), indicating an X-chromosome dosage effect in somatic cells.

The presence of two clonal phenotypes (mutant and wild-type) in females at risk for X-linked diseases has been useful in demonstrating their heterozygosity for the mutation, which is often not possible by examination of uncloned populations of cells (Migeon et. al., 1968; Romeo and Migeon, 1970; Meyer et. al., 1975).

Fate of X chromosomes in germ cells. Analysis of somatic tissues and cultured cells (including fibroblast clones) from human embryos, heterozygous for the G6PDA variant, indicate that a single active X is present in cells from a variety of tissues, at least by five weeks from conception (Migeon and Kennedy, 1975). On the other hand, the presence of the heteropolymorphic form of the enzyme in oocytes of the heterozygous adult female (Gartler et. al., 1972) and those of a 16-week-old fetus (Gartler et. al., 1973) is compelling evidence that, in meiotic stage germ cells, both X chromosomes are active; however, the basis for the two active chromosomes in oocytes is not clear. It is conceivable that germ cell progenitors escape inactivation because it occurs when cells destined to become germ cells have already been imprinted. Alternatively, only a single X chromosome may be expressed in all cells of the early zygote, the second X activated only in term cells at some time following their differentiation. The third possibility is that germ cells are subject to inactivation but reactivated during oocyte differentiation, as suggested by Gartler (1976) and Ohno (1964).

We have studied ovarian extracts from fetuses heterozygous for the G6PDA variant and observed the presence of a heteropolymer as early as eight weeks after conception, at a time when the population of germ cells is almost exclusively pre-meiotic (Migeon and Jelalian, 1977). The presence of two active X chromosomes in pre-meiotic germ cells does not differentiate between escape from inactivation, late activation of the second X, or reactivation of the silent X, but excludes reactivation co-incident with the onset of meiosis.

Time of X-inactivation. There is no direct evidence as to when X-inactivation occurs in any mammal. G6PD variants in man have been used to estimate the size of the embryonic cell population at the time of X-inactivation (Gandini et. al., 1968; Nesbitt and Gartler, 1971). However, the analysis has been complicated by the fact that selection favoring one of the two cell populations resulting from X-inactivation may distort the original mosaicism and even mask it.

II. SELECTION AS A DETERMINANT OF FEMALE PHENOTYPE

In respect to X-linked genes, the phenotype of the female is determined by (1) the extent and nature of her individual *heterozygosity,* (2) the effect of *random inactivation* on the proportions of cells of both types in each tissue, and (3) the result of *selection* following inactivation.

If there were no differences between the maternal and paternal alleles at any X-linked locus, there would be no cellular mosaicism; however, on the basis of estimates of the extent of heterozygosity in human populations (Harris and Hopkinson, 1972), it seems certain that most women are heterozygous for at least one, and probably more than ten X-linked genes.

The presence of unbalanced mosaicism in some females has been interpreted as indicating preferential inactivation of one X or the other; yet, the evidence for directed inactivation in the embryo is not compelling. Skewed distributions of the two clonal populations do not preclude random inactivation because the number of cells destined for various tissues may, at the time of inactivation, be small and subject to the vagaries of chance. In addition, cell migration patterns may obscure the underlying cell heterogeneity. Moreover, there is considerable evidence that selection favoring one cell population occurs after random inactivation. Differences in the nature of maternal and paternal alleles at various X-linked loci in the heterozygote may lead to phenotypic differences in cells which are identical except for the parental origin of their active X chromosome. Variant alleles may confer some proliferative advantage or disadvantage to the embryonic cells in which they are expressed. Competition between the two cell populations resulting from X-inactivation is not only conceivable but likely, and this competition may have a significant role in the determination of the normal female phenotype.

Selection at the level of the chromosome. Selection undoubtedly occurs in somatic cells of the female embryo carrying X chromosomes with altered morphology. Studies of chromosome replication have indicated that structurally abnormal X chromosomes such as deletions, ring chromosomes, and isochromosomes are invariably inactivated in cells in which they occur. This is expected because these variant X chromosomes all have deletions of X chromatin; hence, the cells in which the *normal* X was inactive, being nullisomic for a segment, would not be viable. The alternative explanation, i.e. that the *abnormal* X is in some way recognized as aberrant and preferentially inactivated, is excluded by the replication patterns of X chromosomes involved in translocations with autosomes (Leisti et. al., 1975). In this case, either the normal or variant X may be subject to inactivation. Furthermore, it is the *normal* X which is consistently turned off in carriers of the translocation chromosome who are phenotypically normal; inactivation of the X chromatin of the translocation chromosome is known to turn off genes in adjacent autosomal segments (Cattanach, 1970; Leisti et. al., 1975), which is usually detrimental or even lethal to the cell. Therefore, the best explanation for the non-random inactivation patterns observed with X chromosome variants is cell death, rather than preferential inactivation. Selection of this intensity implies an extensive cell loss in the embryo, yet losses of this magnitude are known to occur during the development of monozygotic twins; similarly, mice of normal size have been derived from partial blastocysts.

Further evidence of selection against cells with unbalanced karyotypes comes from studies of individuals *mosaic* for X chromosome aneuploidy, i.e. XO/XX, whose cells of abnormal karyotype become less frequent as the individual ages (Eller et. al., 1971).

Selection at the level of the gene. Evidence is accumulating that selection also occurs at the level of single genes. Although mosaicism with regard to HGPRT activity is expected in females heterozygous for the Lesch-Nyhan mutation (X-linked HGPRT deficiency), enzyme deficient cells have not been detected among their red blood cells (Kelley et. al., 1969; McDonald and Kelley, 1972). Selection has been implicated as an explanation for the paucity of enzyme deficient cells from studies of females simultaneously heterozygous for two X-linked genes, HGPRT and G6PD (Nyhan et. al., 1970). In this case, the HGPRT wild-type allele was on the same chromosome as that specifying G6PDB. The heterozygous phenotype (G6PDAB) was found in skin fibroblasts, but blood cells expressed only the G6PDB allele, the one coupled with the normal HGPRT allele. The population of cells having the mutant allele and G6PDA on the active X chromosome had been essentially excluded.*

---

*The exclusion of HGPRT⁻ cells is not complete in red cells. Heterozygotes, especially the younger ones, may have minor populations of the disfavored cell (Migeon, unpublished observations).

Selection of this kind also occurs in lymphocytes of the heterozygote, where less than 10% HGPRT deficient cells have been found consistently (Dancis et. al., 1968; Albertini and DeMars, 1968).

It seems likely that cells expressing the Lesch-Nyhan mutation are disfavored because the mutation is associated with a *severe* enzyme deficit. There is no selection against red cells of mutant type from females heterozygous for mutations leading to *partial* deficiency of HGPRT, associated with X-linked gout, and these women may express the mild deficiency (Johnson et. al., 1976).

Selection is tissue specific. Mutations leading to phenotypes disfavored in one tissue may be neutral in another. Analysis of skin fibroblast clones from heterozygous relatives of Lesch-Nyhan males reveals no evidence of a selective advantage of either allele; among 4,000 clones from 22 heterozygotes, approximately half were of the mutant type (Migeon, 1970).

Between individual heterozygotes, however, there is considerable variability; some have a small population of enzyme deficient skin fibroblasts, while others have relatively few clones of normal type. This kind of variability is reflected in the wide range of enzyme activity which characterizes uncloned skin fibroblasts (Fujimoto and Seegmiller, 1970; Migeon, 1971). Although the proportion of HGPRT deficient skin clones varies greatly among heterozygotes, most individuals tested have greater than 10% (Felix and DeMars, 1971; Migeon, 1971). However, it is likely that, in some heterozygotes, the population of cells of mutant phenotype is less than 10%. The sensitivity of heterozygote detection can be increased by devising means to detect rare variant cells among predominantly normal cell populations. At the HGPRT locus, this is done by nutrient medium which favors the growth of mutant cells (Migeon, 1970; Felix and DeMars, 1971). In contrast to cells with normal HGPRT activity, those lacking this enzyme can proliferate in the presence of 6×10^{-5}M 6-thioguanine. Using the ability to proliferate in 6-thioguanine as an assay, we have detected approximately one cell of mutant phenotype among 300 wild-type cells (Migeon, 1977) from an obligate heterozygote for HGPRT deficiency. The low frequency of HGPRT deficient fibroblasts is most likely attributable to some proliferative advantage for the cell with normal HGPRT activity. However, because cells expressing HGPRT deficiency are not usually disfavored in skin, selection in this case most likely is not directed towards the HGPRT locus, but is operating at another locus linked to HGPRT. The paucity of HGPRT deficient cells, therefore is attributable to a hitchhiker effect resulting from selection against an unknown allele coupled with the mutant allele at the HGPRT locus.

III. CELL COMMUNICATION MASKS GENOTYPE AND INFLUENCES SELECTION

That the HGPRT mutation is neutral in skin fibroblasts may indicate that the gene product is not essential to the function of these cells; alternatively, it may be that selection does not occur because of the contact-mediated communication between wild-type and mutant cells. Subak-Sharpe and colleagues (1969) showed that the presence of rare mutant cells can be obscured by contact-mediated communication of the pertinent product from wild-type to mutant cells by "metabolic cooperation". This is not merely a cell culture phenomenon since HGPRT deficient cells are also masked in uncultured skin specimens from obligate heterozygotes (Frost et. al., 1970).

Using an assay which quantitates the communication of nucleotides between skin fibroblasts, we have found that contact feeding of this kind occurs extensively and is a stable characteristic of the cells from all individuals studied (Corsaro and Migeon, 1975). Cell communication mediated through gap junctions (Gilula et. al., 1972) undoubtedly influences the viability of HGPRT deficient cells. Red blood cells and lymphocytes of mutant type, because they do not form gap junctions, maintain a severe enzyme deficiency, and are subject to selection, whereas cells of mutant type in skin bypass the enzyme deficiency via contact transfer of nucleotides. Contact-feeding eliminates differences between mutant and normal cells and therefore any potential proliferative advantage of the normal cells. Gap junction mediated communication is not limited to transfer of nucleotides; it is likely that many ions and molecules of less than 1200 MW are transferred from one cell to another in this way (Simpson et. al., 1976; Corsaro and Migeon, 1977).

Metabolic cooperation of another kind takes place in females heterozygous for X-linked mutations involving lysosomal enzymes. In this case, wild-type cells secrete the enzyme, making it available for uptake by cells expressing the mutation, thereby correcting the defect. In heterozygotes for the Hunter's mutation (X-linked iduronate sulfatase deficiency), as few as 20% skin fibroblasts expressing the normal allele can obscure the presence of the majority of cells expressing the Hunter allele (Migeon et. al., 1977). Correction of this kind obviates the need for selection against cells of mutant phenotype.

It seems certain that the transfer of gene products from one cell to another in some tissues prevents selective overgrowth to a considerable extent, and may explain why the complete elimination of one clonal population in the heterozygous female is not observed more frequently.

IV. THE HETEROZYGOUS PHENOTYPE

As a consequence of random inactivation, one would expect variable expression of the mutant allele among heterozygotes for X-linked genes and that the clinical phenotype of some heterozygotes may be as severe as that of their affected sons. On the other hand, only minimal abnormalities of purine metabolism have been noted among carriers of the Lesch-Nyhan syndrome. Perhaps through metabolic cooperation, the enzyme product is made available to all cells in sufficient quantities to prevent clinical manifestations. It is also conceivable that the proliferation of cells expressing the normal allele is favored not only in blood, but in other pertinent tissues as well.

Selective advantages of the wild-type alleles have been observed at other X-linked loci. Distribution of G6PD red blood cell phenotypes in an unbiased sample of females heterozygous for G6PD Mediterranean mutation (leading to a deficiency of the enzyme) is skewed in favor of the cells expressing the normal enzyme phenotype (Rinaldi et. al., 1976) and there is evidence that the amelioration of clinical symptoms in heterozygotes at the Incontinentia Pigmenti locus may be attributable to selection against the mutant allele (Migeon, unpublished observations).

Advantages of this type, however, would result only if (1) the pertinent locus were expressed in cells at the time they are undergoing proliferation, and (2) the variant influenced the cells ability to proliferate. Conceivably, some mutations might confer a selective advantage.

In any event, it seems likely that X-inactivation occurs randomly and results in two populations of cells differing with regard to their X chromosome phenotype. During embryogenesis, the overgrowth of one cell population may occur with elimination of the other. The results of the competition between the two cell populations may differ from tissue to tissue, depending upon which heterozygous alleles are expressed in that tissue, whether the pertinent gene product influences the growth of a specific type of cell, and whether this product is transferred between cells. Cell communication masks the genotype, corrects the deficiency, and obviates the selective advantage of the wild-type cell.

V. CONTRIBUTIONS TO FEMALE FITNESS

From mortality statistics, it is apparent that the male is more vulnerable than the human female at almost every age. There is an excess of boys among those who die in infancy at the time when environmental differences between the two sexes are minimal (Childs, 1965). The opportunity for females to carry two alleles at the same locus simultaneously must provide versatility not possible in the hemizygous male, masking or ameliorating the effect of the variant allele. Yet, the observations discussed here indicate that selection favoring one of the two cell populations produced by X-inactivation plays a role in determining female phenotype. It seems likely that the competition between cell populations during embryogenesis provides a unique adaptive advantage for the female. Therefore, it is not only the variety of alleles in the female, but also the choice between the two alleles, which contributes to her greater biological fitness.

ACKNOWLEDGMENT

This work was supported by NIH grant #HD 05465.

REFERENCES

Albertini, R. J. and DeMars, R. (1968). *Biochem. Genet.* **11**, 397-411.

Cattanach, B. M. (1970). *Genet. Res.* **16**, 293-301.

Childs, B. (1965). *Pediat.* **35**, 798-812.

Comings, D. E. (1966). *Lancet* **ii**, 1137-1138.

Corsaro, C. M. and Migeon, B. R. (1975). *Exp. Cell Res.* **95**, 39-46.

Corsaro, C. M. and Migeon, B. R. (1977). *Nature* **268**, 737-739.

Dancis, J., Berman, P. H., Jansen, V. and Balis, M. E. (1968). *Life Sci.* **7**, 587-591.

Davidson, R. G., Nitowsky, H. M., and Childs, B. (1963). *Proc. Natl. Acad. Sci. U.S.* **50**, 481-485.

Eller, E., Frankenburg, W., Puck, M. and Robinson, A. (1971). *Pediat.* **47**, 681-687.

Felix, J. S. and DeMars, R. (1971). *J. Lab. Clin. Med.* **27**, 596-604.

Ferguson-Smith, M. A. (1965). *J. Med. Genet.* **2**, 142-155.

Fialkow, P. J. (1976). Biophysics **458**, 283-321.

Frost, P., Weinstein, G. D. and Nyhan, W. L. (1970). JAMA **212**, 316-318.

Fujimoto, W. Y. and Seegmiller, J. E. (1970). *Proc. Natl. Acad. Sci. U.S.* **65**, 577-584.

Gandini, E., Gartler, S. M., Angioni, G., Argiolas, N. and Dell'Acqua, G. (1968). *Proc. Natl. Acad. Sci. U.S.* **61**, 945-948.

Gartler, S. M. (1976). *Fed. Proc.* **35,** 2191-2194.

Gartler, S. M., Liskay, R. M., Campbell, B. K., Sparkes, R. and Gant, N. (1972). *Cell Differentiation* **1,** 215-218.

Gartler, S. M., Liskay, R. M. and Gant, N. (1973). Exp. Cell Res. **82,** 464-466.

Gilula, N. B., Reeves, O. R. and Steinbach, A. (1972). *Nature* **235,** 262-265.

Ham, R. G. and Puck, T. T. (1962). *In: Methods in Enzymology* (eds. Colowick, S. P. and Kaplan, N. O.), Academic Press, New York, Vol. 5, pp. 90-119.

Harris, H. and Hopkinson, D. A. (1972). *Ann. Hum. Genet. Lond.* **36,** 9-20.

Johnson, L. A., Gordon, R. B., and Emmerson, B. T. (1976). *Nature* **264,** 172-174.

Kahan, B. and DeMars, R. (1975). *Proc. Natl. Acad. Sci. U.S.* **72,** 1510-1514.

Kelley, W. N., Greene, M. L., Rosenbloom, F. M., Henderson, J. F. and Seegmiller, J. E. (1969). *Ann. Int. Med.* **70,** 155-206.

Leisti, J. T., Kaback, M. M. and Rimoin, D. L. (1975). *Am. J. of Hum. Genet.* **27,** 441-453.

Linder, D. and Gartler, S. M. (1965). *Science* **150,** 67-69.

Lyon, M. F. (1961). *Nature* **190,** 372-373.

Lyon, M. F. (1972). *Biol. Rev.* **47,** 1-35.

McDonald, J. A. and Kelley, W. N. (1972). *Biochem. Genet.* **6,** 21-26.

McKusick, V. A. (1975). *Mendelian Inheritance in Man* (Catalogs of Autosomal, Dominant, Autosomal Recessive, and X-linked Phenotypes), 4th ed., Johns Hopkins Press.

Meyer, W. J., III, Migeon, B. R. and Migeon, C. J. (1975). *Proc. Natl. Acad. Sci. U.S.* **72,** 1469-1472.

Migeon, B. R. (1970). *Biochem. Genet.* **4,** 377-383.

Migeon, B. R. (1971). *Am. J. Hum. Genet.* **23,** 199-210.

Migeon, B. R. (1972). *Nature* **239,** 87-89.

Migeon, B. R. (1977). *In: Genetic Mechanisms of Sexual Development* (eds. Vallet, L. and Porter, I.) Academic Press, New York, 1977. (in press).

Migeon, B. R. and Jelalian, K. (1977). *Nature* **269,** 242-243.

Migeon, B. R. and Kennedy, J. F. (1975). *Am. J. Hum. Genet.* **27,** 233-239.

Migeon, B. R., Der Kaloustian, V. M., Nyhan, W. L., Young, W. J. and Childs, B. (1968). *Science* **160,** 425-427.

Migeon, B. R., Norum, R. A. and Corsaro, C. M. (1974). *Proc. Natl. Acad. Sci. U.S.* **71,** 937-941.

Migeon, B. R., Sprenkle, J. A., Liebaers, I., Scott, J. F., and Neufeld, E. F. (1977). *Am. J. Hum. Genet.* **29,** 448-454.

Nesbitt, M. N. and Gartler, S. M. (1971). *Am. Rev. Genet.* **5,** 143-162.

Nyhan, W. L., Bakay, B., Connor, J. D., Marks, J. F. and Keele, D. K. (1970). *Proc. Natl. Acad. Sci. U.S.* **65,** 214-218.

Ohno, S. (1964). *In: International Conference on Congenital Malformation,* 2nd, New York, 1963 (papers and discussions compiled and edited by the International Medical Congress, Ltd., New York), National Foundation, pp. 36-40.

Rinaldi, A., Filippi, G. and Siniscalco, M. (1976). *Am. J. of Hum. Genet.* **28,** 496-505.

Romeo, G. and Migeon, B. R. (1970). *Science* **170,** 180-181.

Romeo, G. and Migeon, B. R. (1975). Humangenetick **29,** 165-170.

Russell, L. B. (1961). *Science* **133,** 1795-1803.

Seegmiller, J. E., Rosenbloom, F. M. and Kelley, W. N. (1967). *Science* **155,** 1682-1684.

Simpson, I., Rose, B. and Loewenstein, W. R. (1976). *Science* **195,** 294-296.

Subak Sharpe, H., Burk, R. R. and Pitts, J. D. (1969). *J. Cell Sci.* **4,** 353-367.

Szybalski, W., Szybalska, e. H. and Ragni, G. (1962). *Natl. Cancer Inst. Monogr.* **7,** 75-88.

Development of the Maize Endosperm
as Revealed by Clones

Barbara McClintock

Carnegie Institution of Washington
Cold Spring Harbor Laboratory
Cold Spring Harbor, New York 11724

I. INTRODUCTION

In the course of examining plant and kernel tissues exhibiting clonely expressed phenotypic modifications induced by somatically occurring deletions or by action of a gene-control system, it became increasingly imperative to understand modes of development of these tissues. It was revealed that each well defined sector reflected a clone of cells. It was necessary, nevertheless, to determine whether, in some instances, such sectors arose from a single cell or reflected a segment of a clone, the other segments being displaced in the mature tissues according to dictates of developmental mechanisms. Studies soon revealed that development of some tissues of the plant and also of the kernel does not pass through a single restricted sequence of events. Although the sequences for each tissue operate under a basic developmental theme, variations on this theme do occur and some of them are notable. These variations may introduce distinctive patterns of distribution of the progeny of a single cell. Phenotypic modifications, somatically induced, are able to reveal such thematic variations. Lack of appreciation of the potentials for such variation, and also of a means of detecting them when they occur, can lead to misinterpretations of the origin and significance of sectors and their distributions. In the plant body, for example, spectacular patterns were observed to arise from one such thematic variation.

Although some excellent studies at morphological and cytological levels have outlined basic developmental themes, none can reveal a theme and its variations with the clarity that clonal expressions allow. Steffensen (1968) undertook an extensive investigation of modes of development of the maize plant, utilizing for this purpose clones of cells showing modified gene expressions, each of which was derived from a cell whose genotype had been altered by treatment of the mature kernel with X-rays. From types of patterns produced by progeny cells of one that had been so modified, he was able to recognize not only the basic theme underlying development of the plant body, but also some of its variations. In this symposium, Dr. Coe describes his studies that utilize a similar method but in this instance for the purpose of estimating the number of initial cells that give rise to individual parts of the plant body.

In this report, I shall confine my discussion to the development of the maize endosperm. It is that part of the kernel from which corn starch is derived. With regard to genic composition, the endosperm may be viewed as a "twin" of the embryo that is present in the same kernel. It is, however, a dead-end tissue, serving to feed the germinating embryo rather than contribute tissues to it.

II. STRUCTURE OF THE MATURE KERNEL

Each structure in the mature kernel has a distinctive mode of development that includes a history of successive modifications. For our purposes, the starting point is the floret on an ear. It consists of the membranous floral parts and an ovary having a single ovule within it. From the ovary wall extends a long style, the silk, with numerous hairs protruding from it that serve to hold wind-blown pollen grains that fall upon the silks. Following pollination and fertilization, each ovary develops into a kernel. The florets on an immature ear are arranged in the same order as the kernels derived from them are arranged on the mature ear. During development of the kernel, events are so regulated that each tissue adjusts its developmental progression to coordinate with all others. The endosperm is only one such tissue. These integrations are exceptionally well described and illustrated by Randolph (1936) and briefly outlined and nicely illustrated by Kisselbach (1949) and by Sass (1955), to which the reader is referred for details.

Both the embryo and the endosperm are products of cells located within the embryo sac which, in turn, develops from a single cell within the ovule. This diploid cell undergoes meiosis, producing four haploid spores, three of which disintegrate. The fourth spore enlarges and by several divisions produces an embryo sac (the megagametophyte) having a quite specific organization of its haploid cells. At one end, close to where the pollen tube will enter, is an egg cell and two adjacent cells (the synergids) that partially enclose it. At the opposite pole is a group of cells (the antipodals) whose function is not well understood. They do not contribute, however, to other tissues of the kernel. The largest cell within the embryo sac is known as the central cell. It possesses two haploid nuclei, located close to each other and also close to the egg cell. This large central cell is conspicuously vacuolated, and it plays an important role in the structural organization of the developing endosperm. (For a description of the development of the maize embryo sac, see Cooper (1937). For a description of the contents of the embryo sac at the electron microscope level, see Diboll and Larson, (1966).)

III. POLLINATION, FERTILIZATION AND ENDOSPERM FORMATION

The pollen grains produced in the anthers of the tassel are products of meiotic divisions, each of which gives rise to four haploid spores. Two nuclear divisions within each spore result in a large cell, the pollen grain, having within it a working nucleus (the tube nucleus) and two very small cells with much condensed chromatin in their nuclei. These are the two sperm cells, and they are the product of the second division within the spore. When the pollen is mature the filament of an anther elongates rapidly and the anther is exerted from its floral coverings. The filament is flexible allowing the anther to sway in the wind. A split appears at the tip of each anther lobe and the pollen spills into the air, to be carried by the wind, finally falling to the ground or on some object, or being caught by the hairs of exposed silks of an ear whose ovules have eggs ready for fertilization. Within a few minutes in the latter instance, a tube extends from the pollen grain, penetrates the silk, and descends through the long silk to the ovule where it deposits the two sperm cells within the embryo sac. One sperm fuses with the egg cell, and following fusion of their nuclei, gives rise to the zygote which will sub-sequently develop into an embryo with a large, partially enclosing scutellum (the monocotyledon) that serves as a storage and con-ducting tissue, feeding supplies from the endosperm to the germinating seedling. The nucleus of the second sperm cell fuses with one of the two haploid nuclei within the embryo sac's large central cell. The second haploid nucleus of this cell then combines with this nucleus to produce the primary endosperm nucleus that is triploid.

Within a few hours this triploid nucleus divides. The two nuclei so formed pass to opposite sides of the embryo sac in the vicinity of the zygote. A second nuclear division follows that also results in directed nuclear orientations. The plane of division often must be opposite to that of the first division (see Fig. 9). Each nucleus, in turn, undergoes further divisions, usually in synchrony, until the central cell of the embryo sac is lined with nuclei. At this time, cell walls begin to form about the nuclei and further divisions may commence to fill the free space of the central cell. This filling may be complete or only partial. In the latter instance, the effects may be noted in the mature kernels by distinct spacial demarcations. During the free nuclear division stage, placement of nuclei within the central cell is not random. The products of the four initial nuclei occupy ordered positions. From them the future cells of the endosperm will be derived and these, likewise, follow an ordered plan. The mode of these orderings provides the initial setting for fulfillment of the basic theme of endosperm develop-ment. The ordering may eventuate in symmetrical distributions of the progeny of the four initial nuclei (Fig. 9), in asymmetric distributions that produce asymmetric clones in the mature endosperm, or to striking modifications in disposition of some of the progeny of these nuclei. An illustration of this latter appears in Fig. 6.

Cellular formation within the central cell of the embryo sac is followed by further cell divisions. These are localized mainly in the peripheral zones. They involve the outermost layer of cells and to a lesser but significant extent the cells of the layers below it. As a con-sequence, rows of cells may be observed in the mature endosperm. These may be traced to the position of their initials in the former central cell of the embryo sac. During the period of cell divisions, those cells in the inner regions of the developing endosperm increase in size rather than number. This increase is associated with DNA replications unaccompanied by cell divisions. Those cells that were laid down early in endosperm development become very large, and they are highly polyploid. The large size of these cells and their nuclei may be noted in the photographic illustrations. The younger cells adjacent to them and the still younger ones spreading outward from these latter are not as large nor are they as highly polyploid. The cells close to the outer edge of the endosperm are the last to be formed, and they are relatively small. In short, the size of cells in the mature endosperm that diverge from the former central cell of the embryo sac toward the outer borders of the endosperm becomes progressively smaller with the exception of those cells that are at the base of the endosperm. The progressions are readily observed in the mature endosperms. In the basal segment of the endosperm, however, very large cells funnel from the former central cell of the embryo sac towards the scutellum of the embryo and abut with it at its mid and lower regions. These cells not only are very large, they also are stikingly elongate in the vicinity of the scutellum.

Evidence of the high polyploid nature of the enlarged cells of the mature endosperm may be obtained readily although published accounts that describe events associated with polyploidization are not numerous (Duncan and Ross, 1950; Tschermak-Woess and Enzenberg-Kunz, 1965; Lin, 1977).

In the late stage of endosperm development, the epidermal layer transforms into the aleurone layer. Its cells enlarge and form the special products that are associated with differentiation of this layer. These include the potential for forming anthocyanin pigments that often exhibit brilliant colorations. This potential is not shared, however, by the underlying endosperm cells. They, in turn, have their distinctive products, the most conspicuous of which are the starch granules. During the final stage of endosperm development, cell divisions are confined to those cells that are located in the layers just below the layer of enlarging aleurone cells, and their planes of division may be anti-clinal as well as periclinal. Cell divisions then cease and the cells just formed complete their differentiation by enlarging and by filling their cytoplasms with starch granules. Nevertheless, the cells in these border regions remain small in comparison with many of those that were left behind during the cambium-like growth of the endosperm.

IV. EXAMPLES OF CLONES THAT ILLUSTRATE THE ORIGIN AND ORGANIZATION OF CELLS OF THE ENDOSPERM

The longitudinally split and stained kernels in Fig. 1 will serve to illustrate aspects of endosperm development that were outlined in the previous section. In each half-kernel the endosperm is delimited from

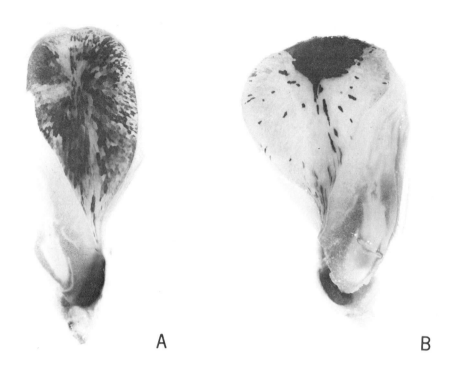

Fig. 1. Halves of two longitudinally cut kernels. Cut surfaces stained with an I-KI solution. A. Embryo area to left sharply demarked from endosperm to right and above. Dark to light staining cells in endosperm reflect differences in amylose content. Note position in endosperm from which rows of cells diverge. The three distinct sectors at upper left come to a focus at this location which represents the initial location from which all cells of the endosperm were derived. Genetic constitution of kernel: standard *a* allele in chromosome 3 and standard *wx* allele in chromsome 9 from ear parent; *a-m3* and *wx-m7* with associated *Ac* from pollen parent. During development of the endosperm the *m* component of *Ac,* initially silent, underwent change to active in some cells. B. Embryo area to right, endosperm to left. The deeply stained cells in endosperm reflect mutational responses at the *wx-m7* locus to the *m* component of *Ac.* The lightly stained cells reflect types of action at this locus when its structure has not been irreversibly altered by responses to signals from the *m* component of *Ac.* Note large sector of darkly stained cells that diverge from a position within the endosperm from which all rows of cells diverge. Constitution of kernel: standard *a* allele and *wx-m7* with associated *Ac* from ear parent; *a-m3* and standard *wx* allele from pollen parent.

the scutellum of the embryo area by a sharp demarcation line. In A, Fig. 1, the embryo area is to the left of the endosperm. The scutellum, adjacent to the endosperm, has channels that enter the mesocotyl of the embryo that separates the shoot in the upper part from the root below it. Individual cells of the endosperm may be distinguished by differences in degree of intensity of stain. The intensities reflect amounts of amylose in the starch granules of a cell, with higher proportions of amylose providing increasing depths of stain.

The gene responsible for differences in staining intensities among the cells is located in the short arm of chromosome 9, there being 10 chromosomes composing the haploid complement of maize. The dominant, wild-type gene is termed non-waxy, and symbolized Wx, in contradistinction to that given to the initially recognized recessive allele. When homozygous, this allele produces endosperms with an appearance resembling candle wax. Thus, it was designated *waxy* with *wx* as its symbol. The wildtype Wx gene is responsible for the production of an enzyme, bound to a membrane of the starch plastid, that replaces amylopectin with amylose (Nelson and Rines, 1962; Nelson and Tsai, 1964). When this Wx gene is homozygous, only a fraction of the amylopectin is changed to amylose. Heterozygotes of Wx with a *wx* allele produce less amylose. Nevertheless, when one dose of Wx is delivered to the primary endosperm nucleus, the amount of amylose that is present in the starch granules of cells of the mature endosperm is sufficient to produce a very intense blue color following application of an alcholic solution of iodine and potassium iodide. Amylopectin, on the other hand, does not stain. Cells having granules with only amylopectin in them show the red-brown color of the stain unless this stain is decolorized by heat rays or removed by hot water, procedures that were followed in these studies.

At different times and in different plants, action of the Wx gene came under the control of either of two well examined regulatory systems, the Activator (Ac) system or the Suppressor-mutator (Spm) system. (For reviews of the action of these control systems, see McClintock, 1965, and Fincham and Sastry, 1974.) The action of other genes have come under the control of one or the other of these systems. Each system is capable of modifying the type and degree of gene action at these various loci during development of either or both the plant tissues or those of the kernel (Dooner & Nelson, 1977). In this report, examples of differences in phenotypic expression that are produced by somatically occurring modifications affecting Wx gene action, and induced by these systems, are made evident by variation in intensity of blue coloration of the starch within individual cells of the mature endosperm when these cells were subjected to the I-KI staining test. As the photographs indicate, these intensities range from faint in some cells to very deep blue in others.

Functioning of the Wx gene and its alleles is limited to the haploid gametophytes (the embryo sac and the pollen grain) and the endosperm of the kernel. It does not function in cells derived from the zygote, represented by the embryo area in kernels, and the diploid cells of the plant. Another genetic system is responsible for production of amylose starch in these plant parts (Akatsuka and Nelson, 1965, 1966). In the illustrations appearing in this report the stained starch in the embryo area of the kernels reflects the functioning of this plant system. The lower concentration of starch in cells of the embryo area as compared with that in the cells of the endosperm is responsible for the reduced density of stain in the cells of this area.

Patterns of cellular distribution within an endosperm are evident among those kernels that have been broken in half longitudinally and stained with an I-KI solution (Figs. 1 to 4 and 6 to 8). Their recognition is facilitated by differences among cells in stain intensity. This allows rows of cells to be traced to their source of origin. Although shapes of endosperms may differ, their patterns of cellular distribution exhibit some common features. It is evident in most instances that the rows of cells come to a focus at one internal location. The focus is often closer to the crown of the kernel than to its base but the exact location depends on the shape of the endosperm which, in turn, may be under genetic control. Nevertheless, the position of focus in each instance refers to the initial endosperm nuclei that lined the central cell of the embryo sac. The progeny of these cells produced the cells of the endosperm, and their alignments away from the original central cell of the embryo sac may be determined in mature kernels. Thus, the basic theme of cell division and cell enlargement during endosperm development is revealed merely by viewing patterns of cell arrangement and cell size within an endosperm.

In Figs. 1, 2, and 3, some well defined sectors (clones) composed of very deeply stained cells, lead directly to the focal location, or can be interpreted to lead there had the plane of split included some cells that were within the full range of the clone. The plane of split sometimes cuts through a segment of a large, clearly marked clone, severing it from the part that leads directly to the focal source in the region of the former embryo sac. Thus, in the illustrations, large clones that are so severed may be referred for their origins to nuclei that lined the embryo sac. In the endosperms of the kernels in Figs. 1 to 3 there are, however, some lighter staining cells whose distributions are patterned but not in a manner that resembles the clones of deep staining cells. From both the genetic and developmental points of view, these odd patterns are highly significant, and their importance needs to be considered.

Differences in expression of the Wx gene among the cells of the endosperms, shown in Figs. 1 to 3, are imposed by the presence at the Wx locus of the controlling element Ac. The modified locus is designated wx-$m7$ as it was the seventh detected instance in my cultures in which action of the Wx gene had come under the control of a known regulatory

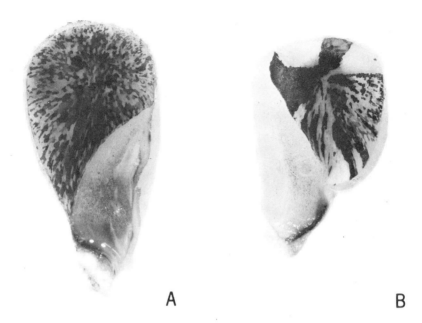

A   B

Fig. 2. Halves of two longitudinally cut kernels. Cut surface stained with an I-KI solution. A. Embryo area to right, endosperm to left and above. Both *wx-m7* and *Ds* (Dissociation, located proximal to *wx-m7* in short arm of chromosome 9) are responsible for the differences in intensity of stain in the cells of the endosperm. Note location within the endosperm from which rows of cells diverge. B. Embryo area to left, endosperm to right. Pericarp layer removed. A potentially active *m* component of *Ac* associated with the *wx-m7* locus was introduced into the primary endosperm nucleus by the ear parent. Note focal location, close to the crown of the kernel, from which diverge two large sectors composed of deeply stained cells and two composed of near colorless cells, and from which all other rows of cells likewise diverge. The plaque of lightly stained cells surrounding the focal area represents one of the distinctive patterns that is produced by *wx-m7* in the cells of this region that have not responded to the *m* component of Ac. Note similar area in B of Fig. 3.

system. The *m* designation in the symbol stands for *Ac*-induced mutations affecting actions of the *Wx* gene that are heritable and usually refractory to subsequent locus modification by the *Ac* element. In the gametophytes or the endosperm such mutations can be expressed in the immediate progeny of one in which such an event occurred. Examples of this are the previously mentioned clones in these figures that are composed solely of cells with deep-staining starch. In addition, however, *Ac* at *wx-m7* induces another class of modulations of the *Wx* gene locus and these are unrelated to those produced by its *m* component. This second class of modulations do not permanently alter the locus. Instead, they represent sequences of "settings" of the locus each of which can effect a modulation that controls the time and degree of action of the

Wx gene in different cells during endosperm development. The times of action may be viewed directly but the degrees are estimated on the basis of staining intensities of the starch within the cells when these cells are subjected to the I-KI staining test. In all instances, however, the patterns this second component of *Ac* induces differ markedly from those produced by the action of the *m* component.

Examples of the second class of *Ac* induced modulations of *Wx* gene expression appear in each of the two longitudinally split and stained kernels in Fig. 4. In both kernels the mutator (*m*) component of *Ac* is completely silent in all cells. Thus, the patterns that are evident among

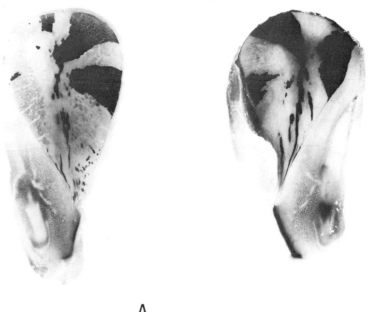

A   B

Fig. 3. Halves of two longitudinally cut kernels. Cut surfaces stained with an I-KI solution. A. Note channels in scutellum of embryo area to left. Endosperm is to right and above the embryo area. The light cracks in that part of the endosperm that is adjacent to the scutellum are due to drying of the moist surface before the kernel was photographed. The ear parent introduced *wx-m7* with its associated *Ac*. Mutations at *wx-m7*, induced by the *m* component of *Ac*, are responsible for the deeply stained cells. Note the sectors composed of such cells that arise from or focus towards the location from which all rows of cells diverge. The near colorless cells surrounding this focal area and the lightly stained cells beyond them, project a pattern that is frequently encountered in endosperms in which the *m* component of *Ac* is silent. See similar pattern in the endosperm of the kernel in A, Fig. 4, and the kernel in Fig. 5. B. Embryo area to right, endosperm to left and above. Pericarp layer removed from crown of kernel. Constitution of the kernel: *wx-m7* with associated *Ac* introduced by the ear parent. This *Ac* initially had a completely silent *m* componenet. The pollen parent introduced an *Ac* at another location in the chromosome complement. It had a potentially active *m* component. Note the mutant sectors composed of deeply stained cells arising from or focusing towards the region of cell row divergence. Note also the pattern produced by differential staining intensities within those cells in which such a mutation had not occurred.

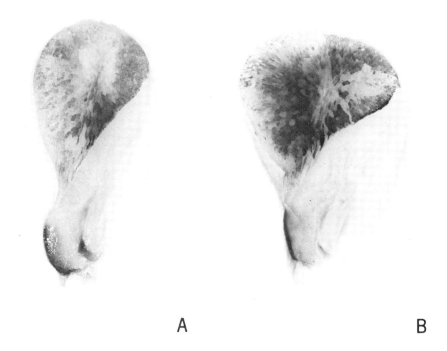

A   B

Fig. 4. Halves of two longitudinally cut kernels. Cut surfaces stained with an I-KI solution. In both kernels the *m* component of the *Ac* that is associated with *wx-m7* was completely silent. The pattern of light and dark stained cells in each endosperm results from a "presetting" of the *wx-m7* locus that is subject to subsequent modification. Note the difference in shape of the endosperm of the two kernels and the relation of this to the position from which all cell rows diverge. In some cells in A of this figure, the nuclei are recognizable as non-stained circular areas within the cells.

the stained cells reflect only the second mode of *Ac* control of gene expression. Its effects appear in those cells that either have reached a particular stage in the development of the endosperm, or occupy spacially reserved regions within the mature endosperm. In A, Fig. 4, there is a group of faintly stained cells about the former location of the central cell of the embryo sac. On each side of this near colorless central region there is an abrupt change to cells that are notably stained. Cells that are similarly stained continue to the outer border of the endosperm. Nevertheless, there are differences in intensity of stain among cells within this border region. Again, some of the enlarged cells that pass from the centrally located clear area toward the base of the endosperm are very deeply stained. Just below this group, however, the cells again are near colorless. The over-all pattern shown in A, Fig. 4, is a common one that has been encountered repeatedly. It shows some degree of heritability through plant generations although no one type of pattern has proved to be stable in the limited tests of this that I have conducted. Some strikingly different patterns do arise. That in B of Fig. 4 is just one example of this.

Patterns provided by this alternate mode of control of Wx gene expression are visible in each of the longitudinally split kernels in Fig. 3. These patterns appear in those parts of the endosperm that do not show responses of wx-m7 to the m component of Ac. Responses to m are evident in those cells that are very deeply stained and are arranged in clones or are expressed only in individual cells.

The mechanism responsible for this alternate mode of programing the expression of the Wx gene has not yet been explored at the molecular level. It is anticipated, nevertheless, that whatever the mechanism may be it could be applicable, generally, in eukaryotic organisms. It would reflect one mode of control of time and type of gene action that does not involve gross modifications in organization of the DNA at the locus. This would be in contrast to the action of the mutator (m) component of Ac that is known to induces changes in DNA composition.

Fig. 5, which appeared in an earlier publication (McClintock, 1965), is reproduced here in order to relate its patterns of gene expression to the theme of this essay. Before the crown region of this kernel was removed, a single, large and well defined area was noted that showed many spots of intense anthocyanin pigment in the aleurone layer. It was known that these spots represented responses of an element that had been inserted at the A (anthocyanin) locus in chromosome 3 to signals from the m component of Ac, and in this instance to the Ac that was located at wx-m7 in chromosome 9. The A gene is one of at least seven whose action is required for production of anthocyanin in the aleurone layer of the endosperm. In a number of instances action of this A gene came under the control of one or another of the known gene-control systems. The one whose action is reflected in Fig. 5 is designated a-m3 because it was the third detected instance in my cultures of such a take-over. The element that had been inserted at the A locus in this instance responds not only to the readily detected action of the mutator (m) component of Ac but also to its non-mutating component, the nature of which has just been described for the Wx locus. It should be emphasized that in most parts of the kernel shown in Fig. 5, the m component of Ac was notably silent. Spots of anthocyanin pigment were confined almost exclusively to the area of the aleurone layer that shows in b of this Figure.

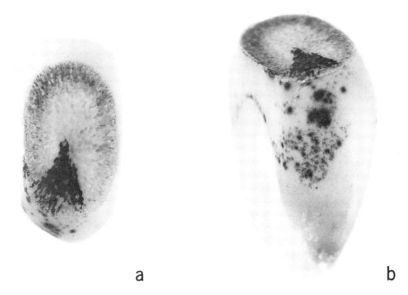

a   b

Fig. 5. Two views of a kernel whose crown region was ground off and the exposed endosperm surface stained with an I-KI solution. From the ear parent the kernel received a-m3 in chromosome 3 and wx-m7 in chromosome 9 whose associated Ac had an m component that initially was completely silent. In a cell, early in endosperm development, it became capable of acting. The V-shaped sector within the endosperm composed of very dark and some lightly stained cells represents the progeny of this cell as does the area with dark spots of anthocyanin pigment in the outermost layer of the endosperm (the aleurone layer). The spots represent mutation-inducing responses at the a-m3 locus to the activated m component of Ac. Note the absence of such spots in areas of the endosperm in which the m component of Ac was silent. Note, also, the different intensities of stain in cells not included within the V-shaped sector, and the pattern these cells present on the exposed surface of the endosperm. (This figure appears in the Brookhaven Symposium in Biology: No. 18 (1965).)

It was postulated that the area in the aleurone layer with pigmented spots represented a clone that had originated from a cell, very early in endosperm development, in which the m component had been activated. If so, this activated m should also have induced mutations at wx-m7, and these should be expressed in some cells of the clone. To verify this postulate, the top of the kernel was ground down to a position that removed the upper part of the area in the aleurone layer that had the prominent spots of anthocyanin pigment. The exposed surface of the endosperm was then stained with an I-KI solution. Its cells gave the differential staining intensities that are readily viewed in a of Fig. 5. The V-shaped sector composed of many deep and some light staining cells in the exposed endosperm surface defines quite clearly the borders of the clone of cells that originated from one in which the m component of Ac had been activated. This clone obviously originated

Fig. 6. Three views of one-half of a longitudinally cut kernel whose cut surface was stained with an I-KI solution. The kernel has *a-m3* in chromosome 3 and *wx-m7* in chromosome 9 with an associated *Ac* whose *m* component produced the pattern of anthocyanin spots and sectors in the aleurone layer shown in a of this figure, and the very darkly stained cells whose sectorial patterns are shown in b of this figure. The circular pattern of anthocyanin sectors in the crown region is matched by a similar circular pattern of deep-stained sectors in the crown region of the inner part of the endosperm, as shown in c of this figure. See text for a discussion of these patterns.

early in development of this endosperm. The focus of the V-shaped sector suggests that activation of the m component was initiated in an early endosperm nucleus, and in one that most likely was located within the former central cell of the embryo sac.

Regulation of Wx gene action by the second, non-mutating component of Ac is expressed in Fig. 5 among all cells of the endosperm in which the m component is silent. The pattern it has produced resembles that shown in A of Fig. 4. The large cells in the exposed central region are lightly stained, whereas those about the rim of the endosperm are more deeply stained with, however, lighter staining cells interspersed among them. An unevenness in this pattern about the rim resembles a similar unevenness in the rim region of the kernel shown in A of Fig. 4.

Fig. 6 is included because it illustrates an interesting and readily appreciated example of deviation on the theme of endosperm development as projected by the previous illustrations. The pigmented areas in the aleurone layer of this kernel (a, Fig. 6) represent responses of a-$m3$ to the m component of the Ac that is located at wx-$m7$, whereas the inner cells of the endosperm that stain deeply with the I-KI solution (b, Fig. 6) represent responses of the wx-$m7$ locus to this same m component. When first observed, the circular arrangement in the crown region of sectors composed of aleurone cells with anthocyanin pigment in them was startling because such an arrangement had not been noted previously. To learn of the developmental origin of this circular arrangement of cells in the outermost layer of this endosperm, the kernel was cut in half longitudinally and the cut surfaces stained with an I-KI solution. Three views of one of these halves are shown in Fig. 6. They reveal that the sectors of deep-staining cells within the endosperm, produced by mutations at wx-$m7$, follow the same circular pattern in the region of the crown. The outermost layer of cells of this region (the aleurone layer) exhibits the series of circularly arranged sectors of anthocyanin pigment that progress sequentially to the tip of the kernel (c, Fig. 6). Below this region, the pattern of cell arrangements resembles that previously illustrated. It should be noted that the rows of cells in the lower part of the endosperm, come to a focus in the region where the circular arrangements commence.

The circular pattern of cell arrangement in the crown region of the kernel in Fig. 6 could be explained if the central cell of the embryo sac had expanded linearly during the stage of free nuclear division, or shortly thereafter, and in the direction of the future crown. The cells in this upper region of the central cell complex could then have undergone divisions diverging from this spike-shaped source and in planes

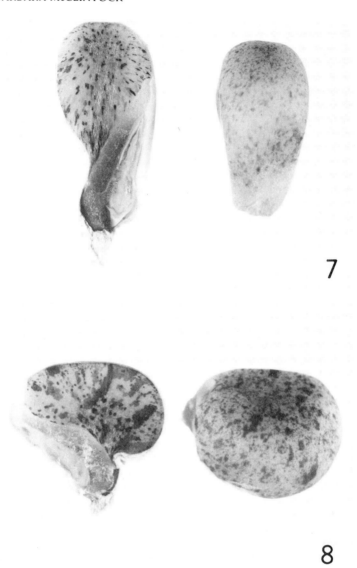

Figs. 7 and 8. Each figure shows both halves of a longitudinally cut kernel. To the left, the cut surface of one of the halves, stained with an I-KI solution, is in view. To the right is the other half of each kernel with its cut surface facing down in order to show the pattern of anthocyanin spots in the aleurone layer of the endosperm. These spots represent responses of *Spm c-m5* to the *m* component of *Spm*. Action of the *C* gene is required for anthocyanin pigment to be made. The dark and lightly staining starch in the endosperm cells of the half to the left represent types of response of *wx-m8* to this *m* component of *Spm*. The kernels received *Spm c-m5* and a standard *wx* allele from the ear parent and a standard *c* allele and wx-m8 from the pollen parent. Note the distinctively different shape of the endosperm in Fig. 7 compared with that in Fig. 8 and that the difference relates to the mode of cellular progression from the initial starting positions.

that verge on the parallel rather than the radial as in the previous illustrations.

The longitudinally cut kernels in Figs. 7 and 8 are included because the pattern of anthocyanin distribution in the aleurone layer of each and the range of Wx gene expressions in their starch-bearing endosperm cells relates to controls by a regulatory system other than Ac. This is the Suppressor-mutator (Spm) system. The Spm in these kernels resides at the locus of the C gene in chromosome 9, and the modified locus is designated c-m5. The presence of Spm at this locus is responsible for inhibition of action of the C gene, and its action is required for anthocyanin pigment formation in the aleurone layer. The m (mutator) component of Spm is responsible for mutations at c-m5 that allow pigment to be produced in those cells of the aleurone layer that are derived from one in which such an m-induced mutation occurred. The Wx locus has an inserted element that responds to signals from the m component of Spm and wherever Spm may be located in the chromosome complement. These responses produce mutations that allow some level of expression of amylose in endosperm cells as determined by the I-KI staining test. Most mutations at this modified Wx locus, designated wx-m8, give rise to stable, heritable modifications of the locus. A number of stable alleles of Wx were isolated that exhibit levels of amylose ranging from low in some isolates to high in others. Although the C and Wx loci reside in the same arm of the same chromosome, they are distantly removed from each other.

In Figs. 7 and 8, both halves of the cut kernels are shown. The exposed cut surface in one of the halves was prepared for the I-KI staining test, and these surfaces are shown to the left of each Figure. The other half of each kernel (right in Figs. 7 and 8) was placed adjacent but in reverse orientation in order to illustrate in a single photograph the reactions of c-m5 and wx-m8 to the m component of Spm.

The aleurone layer of the kernel in Fig. 7 has many closely spaced spots of anthocyanin, and these represent mutations at the locus of c-m5. It is evident, also, that many mutations occurred at the locus of wx-m8. No large mutant clones appear within the exposed parts of the endosperm although mutations were occurring at all stages of endosperm development. The evident mutations are spaced as individual cells or arranged in small clones. This pattern suggests that mutations were not occurring in the leading cambium-like layer of cells but rather in the cells of the layers just below it that also undergo some divisions. This restriction reflects one state of the m component of Spm. Its various states can effect distinctly different but remarkably refined patterns of control of gene expression.

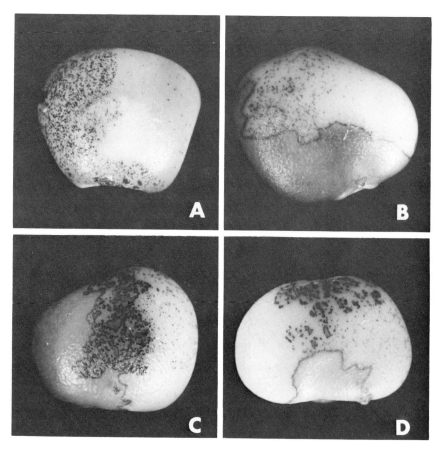

Fig. 9. Top view of four kernels having sharply defined areas in the aleurone layer, each of which expresses a different pattern of pigmentation. The patterns suggest that each area represents a clone of cells that originated from the earliest endosperm nuclei in the embryo sac, and most probably from the first four of these. These photographs appear in volume 16 of the Cold Spring Harbor Symposium in Quantitative Biology where the nature of the events responsible for the pigmented areas are described, (McClintock, 1951).

A quite different state of this *m* component of *Spm* is present in the kernel pictured in Fig. 8. This state is responsible for the pattern of mutations at *c-m5,* made evident by the pigmented spots in the aleurone layer, and also for the mutations at *wx-m8,* registered in the starch-bearing cells of the endosperm. In Fig. 8 (left), a number of clones are visible, each showing a particular level of intensity of amylose stain. It is evident in this instance that mutations at *wx-m8* occurred in cells of the leading cambium-like layer as well as in some cells of the underlying layers. This kernel has an endosperm that is wider than long. The focus of the sectors is not below the crown region as it is in the kernel in Fig. 7 with an endosperm that is longer than wide. Instead, it is close to the scutellum and near the position where the scutellum bulges into the endosperm.

Knowledge of modes of development of the maize endosperm may be obtained in several different ways. The illustrations given here of patterns of action of different genes, each supporting the other in demonstrating the mode of development of a particular kernel, has proven to be effective. Conclusions may be drawn, however, from viewing modified phenotypes that appear in patches over the aleurone layer. Coe (1978) utilized this method and presents his conclusions drawn from it. Examples that provoke conclusions are given in Fig. 9. Nevertheless, in some such instances, developmental origins could be misinterpreted. The kernel shown in Fig. 10 is a case in point. This kernel received c2-m2, located in chromosome 4, and wx-m8, located in chromosome 9, as well as an Spm located elsewhere in the chromosome

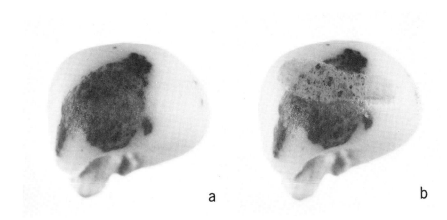

a   b

Fig. 10. Two views of the same kernel. The view in (a) of this figure shows the extent of an area in the aleurone layer of the endosperm having many closely spaced pigmented spots. In (b) of this figure a section of the eleurone layer, passing across the pigmented area, was scraped off and the starch in the underlying cells was stained with an I-KI solution. The kernel has c2-m2 in chromosome 4 and wx-m8 in chromosome 9 both of which respond to Spm wherever it may be located in the chromosome complement. The m component of Spm was in its silent phase when initially introduced into the primary endosperm nucleus. No mutations occur at c2-m2 or at wx-m8 when it is silent. When it is active both loci express mutational events leading to anthocyanin pigment formation in the first instance and to various levels of amylose production in the second instance. The view shown in (b) demonstrates this relationship.

complement. Both c2-m2 and wx-m8 respond to this Spm. In this kernel, the m component of Spm is almost completely silent except in that part of the endosperm that produced the large area in the aleurone layer that has many closely spaced pigmented spots (a, Fig. 10). These represent mutations at c2-m2. That this area reflects activation of the m component of Spm was determined by scraping away a segment of the aleurone layer across and to each side of the heavily spotted area. The underlying starch-bearing cells so exposed, when stained with an I-KI solution, indicated that wx-m8 also had responded to this activated m. Mutations were confined to cells that were within the boundary defined by the overlying aleurone layer having the densely placed pigmented spots. Experience suggests that this large surface area and its underlying cells could represent a single clone initiated by activation of the m component of Spm in a cell very early in endosperm development, and probably in one that was located within the embryo sac complex. Nevertheless, only by making a sequence of sections across this kernel could the origin of the surface sector be analyzed to determine if it truly represents such a single event.

V. CONCLUDING STATEMENT

This report presents a set of photographs to illustrate the basic theme of development of the maize endosperm. The selected examples, however, do not explore the many variations on this theme that should be encountered, especially among kernels produced by some of the races of maize.

In addition, another subject was considered briefly. It concerns a type of control of gene action during development that may apply generally among eukaryotic organisms. Its expression in some of the photographs is so conspicuous that mention of it is unavoidable. It relates to events occurring at the locus of a gene that regulate when the gene will be allowed to function during defined stages of development. In the examples here reported, such events refer to a mode of operation of a known gene-control system. This system prepares a locus in such a manner that its gene will function to a given degree at given stages in development, and it does this without inducing permanent alterations of the DNA at the locus. That such mechanisms must exist in higher organisms has been recognized for may years. At present, however, only postulates are presented in attempts to project the responsible mechanisms at the molecular level. It might be possible to add substance to these postulates by utilizing for analyses materials of the type here described.

REFERENCES

Akatsuka, T. and Nelson, O. E. (1965). *Genetics* **52,** 425-426 (Abstr.).

Akatsuka, T. and Nelson, O. E. (1966). *J. Biol. Chem.* **241,** 2280-2286.

Coe, E. H. (1978). "International Maize Symposium", Urbana 1975 (D. B. Walden, ed.). (In press).

Coe, E. H. (1978). This Symposium.

Cooper, D. C. (1937). *J. Agri. Res.* **55,** 539-551.

Diboll, A. G. and Larson, D. A. (1966). *Am. J. Bot.* **53,** 391-402.

Dooner, H. K., and Nelson, O. E. (1977). Proc. Nat. Acad. Sci. U.S. **74,** 5623-5627.

Duncan, R. E. and Ross, J. G. (1950). *J. Hered.* **41,** 259-263.

Fincham, J. R. S. and Sastry, G. R. K. (1974). *Ann. Rev. Genet.* **8,** 15-50.

Kisselbach, T. A. (1949). *Univ. Nebraska Agri. Exp. Sta. Bull.* **161,** 1-96.

Lin, B-Y. (1977). J. Hered. **68,** 143-149.

McClintock, B. (1951). *Cold Spring Harbor Symp. Quant. Biol.* **16,** 13-47.

McClintock, B. (1965). *Brookhaven Symp. Biol.* **18,** 162-184.

Nelson, O. E. and Rines, H. W. (1962). *Biochem. Biophys. Res. Commun.* **9,** 297-300.

Nelson, O. E. and Tsai, C. Y. (1964). *Science* **145,** 1194-1195.

Randolph, L. F. (1936). *J. Agri. Res.* **53,** 881-916.

Sass, J. E. (1955). *In:* "Corn and Corn Improvement" (G. F. Sprague, ed.), pp. 63-87. Academic Press, New York.

Steffensen, D. M. (1968). *Am. J. Bot.* **55,** 354-369.

Tschermak-Woess, E. and Enzenberg-Kunz, U. (1965). *Planta* **64,** 149-169.

Insertion Mutants and the Control of
Gene Expression in *Drosophila melanogaster*

M. M. Green

Department of Genetics
University of California
Davis, California

I. INTRODUCTION

The genetic control of gene expression in eukaryotes remains one of the major unresolved problems in developmental biology. In contrast to prokaryotes where specific types of control (e.g., repressors in the *lac* operon in *E. coli*) have been identified, little concrete information is on hand for eukaryotes other than the probable identification of a control site in the case of the *rosy* gene in *Drosophila melanogaster* (Chovnick *et al.,* 1976). One possible approach to understanding the genetic basis of control in *D. melanogaster* exploits so-called insertion mutants, mutants caused by the integration of "foreign" DNA into specific genes. This

239

approach is encouraged by the results obtained in the analysis of insertion mutations in prokaryotes, especially *E. coli*. In *E. coli* insertion mutations are caused by at least five different unique DNA sequences designated IS-elements (e.g., IS-1, IS-2, etc.) which occur naturally in the *E. coli* chromosome (Starlinger and Saedler, 1976). If the site of insertion is an operon, polar effects or turning off of genes located distal to the integration site are found. Presumably an equivalent demonstration in a eukaryote would militate for an operon-type organization and regulation.

In the narrative which follows emphasis will be placed on the criteria for identifying insertion mutations in *D. melanogaster* together with a description of a method for the directed production of such mutants.

II. IDENTIFICATION OF INSERTION MUTANTS OF THE *w* LOCUS

In contrast to *E. coli* where five distinctive IS-elements with lengths between 800 and 1,450 nucleotide pairs has been proved by utilizing both genetic and physico-chemical criteria (Starlinger and Saedler, 1976) the occurrence of IS-elements in *D. melanogaster* is inferred from genetic criteria. Thus, in spite of the strong case which can be made for the occurrence of insertion mutations (and thus IS-elements), prudence dictates that until the physical presence of IS-elements is unequivocally established, all such mutations are taken to be presumptive. This is implied in the ensuing discussion. The several genetic criteria which collectively serve to identify insertion mutations in *D. melanogaster* have been considered *in extenso* elsewhere (Green, 1977a). For purposes of discussion here, it is more appropriate to summarize these criteria by briefly describing three different types of insertion mutations localized to the classical white eye (*w*) locus on the X chromosome of *D. melanogaster*.

The three types of insertion mutants are designated as (1) mutable white-crimson (w^c) (Green, 1967), (2) mutable white-deletion (w^{-u}) (Rasmuson, *et al.*, 1974), and (3) mutable wild type (w^{+u}) (Gethmann, 1971). All arose independently of one another, were identified by their inordinately high mutability and were mapped to the *w* locus by conventional genetic methods. Each exhibits distinctive genetic features which will be noted in the synoptic description which follows.

A. Mutable w^c Mutant

The unstable w^c mutant arose following X-ray treatment of the mutant allele w^i (white-ivory). It mutates in both females and males at inordinately high frequencies to wild type (w^+), to *w* and to an array of derivatives whose resultant eye color phenotypes range from near wild type to near white. New mutations occur at a frequency of ca. 1×10^{-3} chromosomes scored. Clusters of phenotypically identical mutants are found when individual w^c males or females are tested for mutability suggesting that many, perhaps all, new mutations occur before meiosis in the germ line. Mutations to wild type are mutationally stable in contrast to the mutants of intermediate eye color phenotype which are as mutationally unstable as is w^c from which they arose. The phenotypically w mutants generated by w^c fall into two classes: one class is mutable just as is w^c; a second class is mutationally stable. The latter, subsequent to detailed cytogenetic analysis, prove to be deletions of varying lengths but with fixed end points. All w deletions begin at the w locus and extend to the right or to the left but do not overlap the w locus. The extent of the segment deleted varies: most deletions are small involving less than one polytene chromosome band. However, some may be so long as to involve the loss of as many as 20 bands. As is the case for new mutants, deletions also occur in either females or males and in clusters suggesting a mitotic rather than meiotic origin. In the course of fine structure mapping of w^c, it became clear that w^c depresses interallelic crossing over. This depression is relieved with the reversion of w^c to w^+. Finally, spontaneous transpositions of w^c to the third chromosome have been recovered. Genetic analysis of these transpositions led to the conclusion that what is specified as the w locus is at a minimum two genetic entities: one structural, one regulatory. The w^c mutation is probably associated with an insertion in the structural component.

B. Mutable w^{-u} Mutant

The presumptive w^{-u} insertion mutation was found on two separate occasions as new, spontaneous white-eye mutations occurring on an X chromosome already containing the white allele, w^{sp} (white-spotted). Since the white-eye mutants are essentially identical in their genetic properties, they will be considered as one. Both mutants exhibited one feature which motivated further genetic analysis: on the basis of phenogenetic criteria each behaved as a deletion; i.e., they · were functionally turned off. Yet their polytene chromosome cytology was normal. Subsequent genetic analysis established two facts consistent with an insertion mutation interpretation. Both reverted spontaneously at a frequency around $1—5 \times 10^{-4}$ chromosomes reinstating the w^{sp} mutant. In fine structure mapping experiments, both w mutants significantly depressed interallelic crossing over. Reversion to w^{sp} was accompanied by a restoration of normal interallelic crossing over. These facts suggest that each w^{-u} is an insertion mutant with polar effects not unlike that described for the *gal* operon in *E. coli*. In each the *w* locus functions as completely "turned off" and thus mimics a deletion. Presumably, the site of insertion in w^{-u} is in the regulatory component of the *w* locus.

C. *Mutable w^{+u} Mutant*

The genetically unstable w^{+u} mutant behaves as a functional w^+ gene. It was identified by its capacity to generate small deletions of the w^+ locus at a frequency of ca. 1×10^{-3} chromosomes. Phenotypically the deletions are male viable and produce a white eye color phenotype. As is the case of the other genetic instabilities, the mutational event is independent of crossing over and is primarily a premeiotic event. Cytologically the deficiencies appear to be normal. In contrast to w^c and w^{-u}, w^{+u} appears to have little or no effect on interallelic crossing over. The absence of a phenotypic effect suggests that w^{+u} is causally associated with an insertion sequence which is inserted between specific subsites of the *w* locus or in an orientation which does not inhibit the "reading" of the *w* locus genetic information. Insertions which do not inhibit "reading" have been described in *E. coli* (Starlinger and Saedler, 1976) and in *D. melanogaster* (Rasmuson and Green, 1974).

D. *About the Nature of the Insertion Mutations*

The foregoing synopsis illustrates the variety of insertion mutations which can occur at one genetic locus. For the moment little can be said about the specific nature of these insertions. What is their physical extent? Is one class of insertion sequence involved in all genetic instabilities or are several different kinds of insertions extant? What is the source of the insertions? Are they normal concomittants of the genome, as in the case of *E. coli,* or are they derived from viral DNA which seems to be ever present in Drosophila cells? In this connection it is important to note that by heteroduplex mapping an apparent insertion sequence has been physically identified in association with the RNA genes of *D. melanogaster* (Pellegrini *et al.,* 1977).

Despite the lack of precise information on the nature of the insertion mutants described, such mutants have the potential for providing material information on the control of gene expression in eukaryotes. If, for example, the operon type organization of *E. coli* has extensive counterparts in a eukaryote such as *D. melanogaster,* a first step in their identification could be the recovery of mutations with polar effects such as have been described for *gal* and *lac* in *E. coli*. Since, by and large, the presumptive insertion mutants reported have occurred sporadically, the recovery of possible polar mutations would be enormously facilitated were there readily available a genetic system capable of generating new insertion mutants at numerous loci in high frequency. In fact, such a genetic system does exist: its source and insertion mutation generating capacity will be described forthwith.

III. THE DIRECTED INDUCTION OF INSERTION MUTATIONS

A. *Mutable sn Mutants*

The search for a genetic system which generates insertion mutants has a serendipitous origin and stems from the discovery of an array of putative insertion mutants at the recessive singed bristle (*sn*) locus on the X chromosome of *D. melanogaster* (Golubovsky *et al.*, 1977). More than a dozen mutable *sn* mutants were recovered, all of which could be traced back to wild type male flies which were collected at several geographically widely separated sites in the U.S.S.R. The wild males were brought into the laboratory, bred and the new *sn* mutants were found among the F_2 male progeny linked to the patroclinous X chromosome. This collection of *sn* mutants exhibited a broad spectrum of phenotypes from slight to extreme departures from wild type not unlike that already described for spontaneous and mutagen induced mutants at the *sn* locus (Lindsley and Grell, 1968). They did, however, differ in their inordinately high mutational instability with each mutant reverting to wild type at a frequency of around $1 \times 10^{-3} - 10^{-4}$ chromosomes scored. In addition, some *sn* mutants mutated to a phenotypically extreme *sn* type which was judged to be a deletion. All in all the *sn* mutants behaved as insertion mutants according to the genetic criteria outlined above.

B. *The MR Chromosome Mutator*

The comparatively frequent occurrence of the mutable *sn* mutants in contrast to the sporadic occurrence of other unstable mutants in *D. melanogaster* raised this crucial question: What is the origin of the mutable *sn* mutants? Because they occurred among the progeny of wild flies collected at widely separated sites, it seemed more reasonable to seek a genetic rather than environmental cause for their origin. Among the possible genetic causes, a mutator gene, widespread in wild populations of flies, represented an attractive hypothesis, especially since mutator genes have already been extracted from wild flies (cf. summary in Green, 1976). This hypothesis became all the more attractive with the realization that a naturally occurring, geographically cosmopolitan mutator gene(s) has already been described in *D. melanogaster*. This mutator is the so-called male recombination (*MR*) second chromosome originally described by Hiraizumi (1971) and subsequently studied in several different laboratories. (For a summary cf. Woodruff and Thompson, 1977.) As originally described by Hiraizumi, *MR* second chromosomes appreciably increase the frequency of mitotic crossing over in males. Subsequently, it was noted that in males *MR* chromosomes increase the frequency of chromosome aberrations and, as measured by the frequency of recessive lethal X chromosome mutations, act as potent mutator genes. Furthermore, *MR* chromosomes occur in high frequency (20-50% of all second chromosomes) in wild populations of *D. melanogaster* and have been found wherever sought; e.g., at numerous sites in the U.S., and in Australia, Greece, Japan, Yugoslavia, England, Taiwan and Israel. The frequency and distribution of *MR* chromosomes made it highly probable that *MR* is also widespread in the U.S.S.R. Thus, one hypothesis for the origin of the unstable *sn* mutants which can be tested is that they originate from *MR* chromosomes present in flies collected in the wild and bred in the laboratory.

C. *Evidence for MR Chromosome Induced Insertion Mutations*

Since no *MR* chromosome of U.S.S.R. origin was available for study, two *MR* second chromosomes isolated from wild flies collected in widely separated geographic sites (viz., Haifa, Israel and Napa county, California) were studied. These chromosomes, designated *MR-h12* and *MR-102*, both increase male recombination and act as mutators as defined by the X-linked lethal test. The ability of *MR* chromosomes to produce insertion mutations, especially *sn* mutants, was carried out in a simple, direct way. [This has been described elsewhere and need not be repeated in detail here (Green, 1977b).] Males carrying either *MR-h12* or *MR-102* and a wild type X chromosome were crossed to females whose X chromosomes consisted of *Maxy*, a chromosome carrying 13 different recessive mutants all easily identifiable on the basis of their homozygous phenotype and a balancer X chromosome. (*Maxy* X chromosomes are viable only as heterozygotes.) New mutants were scored in F_1 females carrying the paternal wild type X and maternal *Maxy* X chromosomes.

The results of the mutation experiments are given in Table I. The control mutation frequencies are taken from Schalet (1957 and personal communication) who used the same *Maxy* chromosome and a laboratory wild type stock. A cursory examination of the data in Table I reveals that, compared to the control, each *MR* chromosome caused a substantial increase in the number of new mutants at the *sn*, *ras* (*ras* = raspberry eye color) and possibly *y* (*y* = yellow body color) loci. The magnitude of the increase becomes more apparent when the mutation data in Table I are recalculated on the basis of mutation frequency per 10^5 chromosomes scored. This has been done for the *y*, *w*, *sn* and *ras* loci and are given in Table II. The data in Table II show that the *MR* chromosomes brought about a modest 4-5 fold increase in mutation at the *y* locus, striking 100-350 fold increases at the *sn* locus, 70-80 fold increases at the *ras* locus and no increase at the *w* locus.

TABLE I

Mutation Induction at Specific Loci by MR

Source MR	Locus and number of mutants recovered*													chromosomes scored
	y	pn	w	rb	cm	ct	sn	ras	v	m	g	f	car	
h12	4	0	1	0	1	0	26	12	0	0	1	0	0	35,455
102	4	1	0	0	1	0	11	13	0	1	1	0	0	47,354
Control	5	1	5	0	4	9	1	2	0	1	5	3	1	490,000

*For an explanation of the gene symbols cf. Lindsley and Grell (1968).

By way of determining whether or not the *MR* induced mutants are presumptive insertion mutants, the mutational instability of a sample of mutants at each of the *y*, *sn* and *ras* has been determined. This has been considered elsewhere in detail (Green, 1977b). It will suffice here to note that among 3 *y*, 6 *sn* and 3 *ras* mutants tested, each proved to be highly revertable implying each to be an insertion mutant. Accordingly, it is reasonable to conclude that *MR* chromosomes provide a method for generating insertion mutants. Thus, in principle, it should be possible to get at the question of regulation of gene expression by exploiting the propensity of *MR* chromosomes for making insertion mutants. As suggested earlier by recovering such mutants at gene loci whose polypeptide product(s) is known, insights into the nature of regulation should be possible.

TABLE II

Frequency of Mutation per 10^5 Chromosomes at Selected Loci

Source of Mutants	Locus			
	y	w	sn	ras
h12	5.6	1.4	73	34
102	4.2	0	23	27
Control	1.0	1.0	0.2	0.4

IV. CONCLUDING COMMENT

The mutator action of *MR* chromosomes described here raises additional questions, some of which were noted earlier. Why are some gene loci mutated, others not? Does this imply some kind of undefined recognition? What is the source of the inserted DNA? How big is the inserted DNA segment? Is there more than one size class of insertions? Are they analogous to the IS sequence of *E. coli* which are normal components of the genome or to the bacteriophage *mu* which integrates at various sites in the *coli* genome?

For the present these questions cannot be answered, but they do open a fertile field for research in the immediate future.

ACKNOWLEDGMENT

The author's research is supported by U.S.P.H.S. grant GM 22221.

REFERENCES

Chovnick, A., Gelbart, W., McCarron, M., Osmond, B., Candido, E.P.M. and Baillie, D. L. (1976). *Genetics* **84**, 233-255.

Gethmann, R. C. (1971). *Molec. Gen. Genet.* **114**, 144-155.

Golubovsky, M. D., Yu, N. Ivanov and Green, M. M. (1977). *Proc. Nat. Acad. Sci. U.S.* **74**, 2978-2980.

Green, M. M. (1967). *Genetics* **56**, 467-482.

Green, M. M. (1976). *In:* "The Genetics and Biology of Drosophila" (M. Ashburner and E. Novitski, eds.) Vol. 1b, pp. 929-946. Academic Press, New York.

Green, M. M. (1977a). *In:* "DNA Insertion Elements, Plasmids and Episomes" (Bukhari, A. L., Shapiro, J. A. and Adhya, S., eds.) Cold Spring Harbor Laboratory, Cold Spring Harbor pp. 437-445.

Green, M. M. (1977b). *Proc. Nat. Acad. Sci. U.S.* **74**, 3490-3493.

Hiraizumi, Y. (1971). *Proc. Nat. Acad. Sci. U.S.* **68**, 268-270.

Lindsley, D. L. and Grell, E. H. (1968). "Genetic Variations of Drosophila" Carnegie Inst. Wash. pub. 627.

Pellegrini, M., Manning, J. and Davidson, N. (1977). *Cell* **10**, 213-224.

Rasmuson, B. and Green, M. M. (1974). *Molec. Gen. Genet.* **133**, 249-260.

Ramuson, B., Green, M. M. and Karlsson, B.-M. (1974). *Molec. Gen. Genet.* **133**, 237-247.

Schalet, A. (1957). *Genetics* **42**, 393.

Starlinger, P. and Saedler, H. (1976). *Current Topics Microbiol. Immun.* **75**, 111-152.

Woodruff, R. C. and Thompson, J. N. (1977). *Heredity* **38**, 291-307.

Index